SpringerBriefs in Ethics

More information about this series at http://www.springer.com/series/10184

Bashir Jiwani

Clinical Ethics Consultation: A Practical Guide

 Springer

Bashir Jiwani
Fraser Health Ethics Services
   and Diversity Services
Burnaby, BC, Canada

ISSN 2211-8101   ISSN 2211-811X (electronic)
SpringerBriefs in Ethics
ISBN 978-3-319-60374-2   ISBN 978-3-319-60376-6 (eBook)
DOI 10.1007/978-3-319-60376-6

Library of Congress Control Number: 2017946846

© The Author(s) 2017
This work is subject to copyright. All rights are reserved by the Publisher, whether the whole or part of the material is concerned, specifically the rights of translation, reprinting, reuse of illustrations, recitation, broadcasting, reproduction on microfilms or in any other physical way, and transmission or information storage and retrieval, electronic adaptation, computer software, or by similar or dissimilar methodology now known or hereafter developed.
The use of general descriptive names, registered names, trademarks, service marks, etc. in this publication does not imply, even in the absence of a specific statement, that such names are exempt from the relevant protective laws and regulations and therefore free for general use.
The publisher, the authors and the editors are safe to assume that the advice and information in this book are believed to be true and accurate at the date of publication. Neither the publisher nor the authors or the editors give a warranty, express or implied, with respect to the material contained herein or for any errors or omissions that may have been made. The publisher remains neutral with regard to jurisdictional claims in published maps and institutional affiliations.

Printed on acid-free paper

This Springer imprint is published by Springer Nature
The registered company is Springer International Publishing AG
The registered company address is: Gewerbestrasse 11, 6330 Cham, Switzerland

Background

My involvement in the world of ethics consultation began in the mid-1990s when I had the good fortune of working with some thoughtful and compassionate ethics professionals: notably, Michael Burgess, Michael McDonald, Alister Browne, John Dossetor and Paul Byrne. There was no formal process map to guide us (them, really) in this practice, no resource manual that suggested whom to meet first or how to conduct a meeting. These experienced individuals just seemed to know what to do, and it was up to me to learn as best I could by following their lead.

In the late 1990s, I moved on to work with various ethics committees in Alberta. Many of these were hospital-based committees, while others served newly formed health regions. These committees were responsible for providing clinical ethics consultations but had been given little formal training on what these consultations should involve. Ethics committees were new to the province, and members approached the Provincial Health Ethics Network (PHEN) of Alberta for help with establishing, providing, resourcing, and evaluating services. Work on committees, and later with PHEN, led me to develop educational resources on ethics consultation, building on the work of my mentors and others in the field and on the growing literature.

What became clear to me was that the success of ethics interventions had so far depended on the personal skills of ethics consultants, together with their sincere and genuine concern for the respectful treatment of people. But what about the rest of us who could not rely on exceptional personal qualities or inherent skills? Exhortations to "be like them" would not be useful unless we could see what the exemplars were up to. So began my project of trying to articulate the elements and processes of ethics consultation.

The ensuing pages offer, in accessible language and format, my understanding of both the practice and implications of good ethics consultation. The book is based on my understanding of debates within the ethics consultation literature and on my experiences with the Vancouver Hospital and Health Sciences Centre Ethics Committee, PHEN, numerous ethics committees in Alberta, a number of national committees, and Providence Health Care and Fraser Health in British Columbia.

How one thinks about the professional practice of ethics will have important implications for questions of professional accountability, including appropriate training and resource requirements for undertaking this work. In a continually changing health-care context, providing ethics consultation is an especially complicated endeavour in which we are all learners. The approach I offer here is not uncontroversial. Other ideas about the goals of ethics consultation and what makes a decision ethically justified exist. What's more, I expect that some of my own views presented here will likely change over time. My hope is that this guidebook will help those who, like me, are attempting to provide effective ethics consultation in health care.

Bashir Jiwani

Contents

1	**Ethics Consultation**	1
1.1	Goals of Ethics Consultation	1
1.2	Living with Integrity	2
1.3	Key Assumptions of This Clinical Ethics Consultation Method	10
	1.3.1 Knowledge Is Situated and Comes from Engagement with Difference	10
	1.3.2 Deliberation Grounded in Trust and Respect Is Necessary to Live with Integrity	12
	1.3.3 Ethics Is the Journey Towards More Ethically Justified Perspectives and Solutions	15
1.4	The Ethics Consultant	16
1.5	Types of Consult Requests	20
1.6	Types of Consult Support	21
1.7	Models of Ethics Consultation Services	25
References		26
2	**The Process of Ethics Consultation**	29
2.1	Consult Stage 1: Pre-consult	30
2.2	Consult Stage 2: Interviews	33
2.3	Consult Stage 3: Mid-consult	41
	2.3.1 Initial Ethics Analysis	41
	2.3.2 Planning Next Steps	47
2.4	Consult Stage 4: The Consult Meeting	48
	2.4.1 Conducting the Consult Meeting	50
2.5	Consult Stage 5: Post-consult	55
	2.5.1 Documentation	55
	2.5.2 Follow-up Support Plan	56
	2.5.3 Evaluating the Ethics Consult	56
	2.5.4 Identifying Systemic Issues	57
2.6	Conclusion	58

Introduction

As an ethicist, I encounter people from many different areas within the health system and professional backgrounds. When discussing issues with me, sometimes colleagues say things like, "Bashir, this isn't really an ethical issue; it's a finance issue" (or a clinical practice issue or a communications issue and so on). Almost all areas implicitly privilege certain values. For example, medicine is fundamentally about advancing patient wellbeing. The law is fundamentally about adherence to rights and responsibilities. In any area of activity, multiple important things must be weighed against the central values. Further, the central values themselves need to be specified in relation to ever-changing and varying contexts. In my view, ethics is not a discrete area of activity. Rather, all decisions involve ethics. *Ethics is not an area or compartment of life; it is a way of looking at life – a lens through which to see any and all of our experiences.* The real question is not whether an issue is ethical, but whether the position taken in response to an issue is ethically justified.

Ethics consultation is a practice through which the quality of ethical justification of a response to an issue can be improved and even maximized so that those affected can live with greater integrity. The ethical justification of a shared decision, action, or attitude depends in part on how accurately and comprehensively contexts are understood and how carefully relevant values have been thought through. The quality of our thinking depends on how inclusive, respectful, and deliberative the process of analysis and decision-making is, as well as how open we are to ambiguity and differences in interpretation.

In health care, decisions are often made in institutional environments that are acutely unequal on multiple levels. People who don't know one another, and who may have different values, are brought together to make life and death decisions. Grave decisions must sometimes be made quickly and in contexts of factual uncertainty. Patients and family members, professionals with differing concerns and loyalties, administrators with constraints and mandates, leaders with social and political concerns, and so on are forced to collaborate to resolve questions. While most societies today include a public that is diverse and significantly unequal, these differences are considerably heightened in the health-care setting.

Health-care professionals are frequently aware that decision-making has an important ethics dimension. They know this at gut level; indeed, practitioners experience visceral reactions, tension, and anxiety when working through ethically complex decisions, now commonly understood as symptoms of moral distress (Rodney 2012; Webster and Baylis 2000). Additionally, decision-makers do not come with fully developed understandings of the facts, values, or meanings of unfolding events. Instead, they come with partial, situated perspectives that can and do need to be developed in order for them to make meaning of situations and then respond in ways that are ethically justifiable – from their own perspectives.

It is now recognized that modern health systems require a correspondingly sophisticated support structure for decision-making. An increasingly important part of this structure is an ethics consultation service. Health-care systems are turning more and more to professional ethicists and to internal groups of ethics consultants, who bring the concepts and methods that improve existing deliberation and decision processes. Ethics services can be especially valuable for particularly significant decisions and decisions marked by uncertainty, disagreement, or controversy.[1]

This book describes one approach to the practice of ethics consultation and offers a clinical ethics consultation process.[2] The purpose of the book is to describe, rather than defend, the model on offer. The intended audience is practitioners of ethics consultation who are looking for a clear and well-developed method for their practice. Here in the introduction I provide some rationale for the approach.

The practice of ethics consultation is still quite young and is set within the broader, emerging field of bioethics. Bioethics is heavily influenced by moral philosophy: the development and critique of moral theories aimed at unveiling the truth about the nature of the world, what a meaningful life involves, and what is important in human conduct. A moral theory is a conceptual structure, based on a set of assumptions about the world, which aims to categorize various aspects of human character, behaviour and relationships and offers standards for evaluating these (Jamieson 1991).

Different moral theories have different approaches to justification. Foundational approaches begin with basic beliefs and then construct consistent and defensible theories that logically flow from these basic premises. The basic beliefs are understood as either self-evident or directly justified by experience (Harman 2000). Examples of foundational approaches include consequence-based models such as Utilitarianism, which emphasizes outcomes (Goodin 1995) and obligation-based theory, such as Kantianism, which emphasizes the rightness of actions themselves (O'Neill 1975). The simple application of moral theory to practical problems is recognized as an insufficient method of analysis and decision-making in the real

---

[1] See Frolic et al. 2012 for a good summary of the variety of ethics practitioners in the Canadian health care context and the journey towards professionalization of this service in Canada.

[2] Useful comparators of this approach can be found in *Ethics Consultation: Responding to Ethics Concerns in Health Care*, which is part of the Integrated Ethics Initiative of the Veterans Health Administration in the US and *Ethics Consultation: A Practical Guide* by La Puma and Schiedermayer 1994. An especially helpful guide early in my practice was *Health Care Ethics Committees: The Next Generation* by Wilson Ross et al. 1993.

world for many reasons, including the alien nature of moral theories to most people, the diversity of reasonable basic beliefs that lead to different theories (so we can't easily decide which theory is the right one), and the reasonable diversity of interpretations of any given theory (even if we agree which theory is the right one, differences in our application of the theory remain).

Bioethics, and ethics consultation, has favoured approaches that suggest basic beliefs about good decisions must be justified by their relationship to other beliefs and also must be grounded in everyday moral experience. The most commonly accepted method of ethical justification in bioethics today is reflective equilibrium (Rawls 1971; Daniels 1979; Strong 2010). This approach suggests that in trying to determine what is ethically justified, we start by identifying our well-considered beliefs, based on experience. From these basic beliefs, we articulate broader principles that account for these beliefs and then continue to check these principles against our considered judgments. As we continue this process of checking up and down, we achieve reflective equilibrium.

The field of bioethics and the practice of ethics consultation has been and remains heavily influenced by the principle-based approach offered by Beauchamp and Childress (2013) in *The Principles of Biomedical Ethics* (now in its seventh edition). Their principle-based approach, which aligns with the method of reflective equilibrium, suggests that there are four basic principles that should inform the analysis of any issue or question: respect for autonomy, justice, beneficence, and non-maleficence. These principles are informed by influential moral theories and cohere with intuitive practical judgments in real-life situations.

The articulation and specification of principles is important because it clarifies the various value alternatives that might guide practical decisions (Martin and Singer 2003). For example, in trying to determine what treatment to offer a dying patient, discussing, engaging, and testing specific principles can help clarify meanings about the value of life, quality of life, and beliefs about death and dying. Principle-based approaches, such as Beauchamp and Childress' model, effectively push the tension between foundational beliefs one step closer to the ground, requiring users to ask which principles should receive priority and how they should be interpreted in real-life contexts. The directive nature and lack of operational clarity of specific principles can be seen as both limiting (why am I restricted to these principles and how do I balance them?) and advantageous (they are great starting points and leave room for contextual interpretation). In my view, one of the key challenges with this approach at the level of ethics practice is that the principles do not resonate well enough with the actual language and concerns of the diverse people involved at either the individual case- or system-level of decision-making. Together with the fact that starting with principles precludes discussion of actual concerns, I have found that taking a principle-based approach confounds rather than supports ethics analysis in real-life situations.

A deeper worry relates to the thinning of justification that comes with principle-based approaches. Taylor (1991) describes this as the rise of instrumental reason where the attention to solving problems focuses on the most expedient way forward without concern for the "sacred structure" of society. MacIntyre (2007) raises a similar concern in terms of the absence of a unified way of understanding human life, human virtue, and what it means to be a good human being. Both recognize that the moral theories on which principle-driven approaches are based are functions of modernity, where discussions of morality have become unmoored from deeper notions of the good life and the connection of the individual to the community. One consequence is that moral perspectives grounded in comprehensive value systems that do not ascribe to the utilitarian concepts and language of modern moral principles are undermined (e.g. see Shah-Khazemi 2007).

A compelling related set of concerns is categorized well by Walker (2008). She argues that for an approach to bioethics to be sound, it must attend to four important themes: (1) how we understand what it means to be a person who has values and is able to make meaning in life, (2) the complex nature of the relationships within which we find our sense of belonging and meaning and negotiate life's challenges, (3) the inherent power dynamics within these relationships, and (4) the inescapable situatedness of the ethics expert, who cannot help but see the situation from a certain social, cultural, linguistic, normative perspective and does not occupy some decontextualized, ideal vantage point.

The approach to ethics consultation I offer in this book can be seen within the coherentist theoretical tradition in that it seeks to balance considered judgments about what is important in actual situations with deeper commitments. It begins at the ground level with inductive processes that eventually lead to the articulation of value themes. It attends to the relational subtleties in these conversations, including power dynamics, incomplete participant perspectives, and the bioethicist's role as a situated resource, helping those involved to deepen and broaden their understandings of contexts and what should matter. Together, participants will ideally emerge with overlapping, if not shared, moral horizons as they determine a way forward. Most importantly, this process seeks to make room for conversations about the good life and what it means to be a good human being as those involved pursue the right thing to do.

My hope is that the following set of conceptual and practical tools will assist those participating in the practice of ethics consultation to critically reflect on and improve their practice. Accordingly, the text that follows is divided into two sections. First, I lay out the thinking behind the approach. I begin by discussing the goals of ethics consultation and introducing several key assumptions and concepts on which my *Clinical Ethics Consultation* approach is based. This section introduces the language and values that are central to this consultation process. I also address the role of the ethics consultant, types of consult requests and supports available in health care, and various models of ethics consultation.

In the second section, I provide a step-by-step model for how to go about each of the five stages of the *Clinical Ethics Consultation* process: Pre-consult, Interviews, Mid-consult, Consult Meeting, and Post-consult. This book also has a companion piece, *The Clinical Ethics Consultation Toolkit*, which offers practical tools for each phase of the consultation process, as well as many sample forms and worksheets useful for ethics consultants and consultant teams. The second section of this guide refers to the *Toolkit* at each of the five stages of the consultation process and describes the templates and worksheets available therein.

References

Beauchamp, Tom L., and James F. Childress. 2013. *Principles of Biomedical Ethics*. 7th ed. New York: Oxford University Press.

Daniels, Norman. 1979. Wide Reflective Equilibrium and Theory Acceptance in Ethics. *The Journal of Philosophy* 76 (5): 256–282.

Fox, Ellen, Kenneth A. Berkowitz, Barbara L. Chanko, and Tia Powell. 2006. *Ethics Consultation: Responding to Ethics Questions in Health Care*. National Center for Ethics in Health Care. http://www.ethics.va.gov/ECprimer.pdf. Accessed 29 Feb 2016.

Frolic, Andrea, and the Practicing Healthcare Ethicists Exploring Professionalization (PHEEP) Steering Committee. 2012. Grassroots Origins, National Engagement: Exploring the Professionalization of Practicing Healthcare Ethicists in Canada. *HEC Forum* 24 (3): 153–164.

Goodin, Robert. 1995. *Utilitarianism as a Public Philosophy*. New York: Cambridge University Press.

Harman, Gilbert. 2000. *Explaining Value and Other Essays in Moral Philosophy*. New York: Oxford University Press.

Jamieson, Dale. 1991. Method and Moral Theory. In *A Companion to Ethics*, ed. Peter Singer, 476–487. Oxford: Blackwell.

La Puma, John, and David Schiedermayer. 1994. *Ethics Consultation: A Practical Guide*. Boston: Jones & Bartlett Publishers.

MacIntyre, Alasdair. 2007. *After Virtue: A Study in Moral Theory*. 3rd ed. Indiana: University of Notre Dame Press.

Martin, D., and Peter Singer. 2003. A Strategy to Improve Priority Setting in Health Care Institutions. *Health Care Analysis* 11 (1): 59–68.

O'Neill, Onora. 1975. *Acting on Principle: An Essay on Kantian Ethics*. New York: Columbia University Press.

Rawls, John. 1971. *A Theory of Justice*. Cambridge: Harvard University Press.

Rodney, Patricia. 2012. Moral Agency: Relational Connections and Trust. In *Toward a Moral Horizon*, ed. Janet Storch, Patricia Rodney, and Rosalie Starzomski, 2nd ed., 153–177. Toronto: Pearson Education Canada.

Shah-Khazemi, Reza. 2007. *Justice and Remembrance: Introducing the Spirituality of Imam Ali*. London: I.B Tauris and Company.

Strong, Carson. 2010. Theoretical and Practical Problems with Wide Reflective Equilibrium in Bioethics. *Theoretical Medicine and Bioethics* 31 (2): 123–140.

Taylor, Charles. 1991. *The Ethics of Authenticity*. Boston: Harvard University Press.

Walker, Margaret Urban. 2008. Groningen Naturalism in Bioethics. In *Naturalized Bioethics: Toward Responsible Knowing and Practice*, ed. Hilde Lindemann, Marian Verkerk, and Margaret Urban Walker, 1–20. Cambridge: Cambridge University Press.

Webster, George, and Francoise Baylis. 2000. Moral Residue. In *Margin of Error: The Ethics of Mistakes in the Practice of Medicine*, ed. Susan B. Rubin and Laurie Zoloth, 217–230. Hagerstown: University Publishing Group.

Wilson Ross, Judith, John W. Glaser, Dorothy Rasinski-Gregory, Joan McIver Gibson, and Corrine Bayley. 1993. *Health Care Ethics Committees: The Next Generation*. Chicago: Wiley Publishing.

Chapter 1
Ethics Consultation

Ethics consultation uses skills and knowledge from the traditions of ethics analysis and dispute resolution to help individuals, teams and organizations develop ethically justified responses to challenging situations. Challenges arise when there is disagreement or uncertainty about the best way forward in the situation. These challenges are heightened by the strong emotions and relational conflicts the people involved are often experiencing. An *ethically justified response* occurs when the people primarily affected by an issue arrive at a shared understanding of the facts that underpin a situation and the values that should guide the decision or response, and then reach a solution that aligns with these facts and values. When this happens those involved experience greater integrity. Ethics consultants support those involved to treat each other with respect through a deliberative process that allows more ethically justified responses to emerge.

1.1 Goals of Ethics Consultation

Some of the goals commonly identified for ethics consultation include: finding an ethical resolution to the problem, improving relationships between those involved, building capacity to make better decisions in the future, promoting high quality health care practice and identifying troublesome patterns that systemic change can improve (Andre 1997).

The immediate goal of ethics consultation is to help people achieve *ethically justified decisions in the situation at hand*. This means decisions that: (a) are based on an inclusive and respectful deliberative process that involves all interested parties in a balanced way; (b) include a thorough review to identify the best available information about the situation (what I will call facts); and (c) incorporate a careful, systematic and well-rounded consideration of what should matter most in the situation to everyone involved (values), with particular attention to the values of those to whom a fiduciary duty is owed, and keeping in mind the broader value commitments of society.

Ethics consultation is a resource to help (prospective) decision-making in challenging situations or (retrospective) meaning-making for difficult decisions that have already been taken. This practice is grounded in the values of integrity, trust and respect. Ethics consultation involves a process that enables those involved to make decisions or reflect on past decisions through formal and systematic methods for understanding and evaluating facts, values and emotions. The process creates formal space for including the perspectives of patients, family members and others affected by the decision.

In the context of individual patient care, the goal of ethics consultation is often to arrive at care and treatment plans. At the system level, ethics consultation may be called upon in setting informal and formal policies, practice guidelines and strategies. In almost all cases, those working within the system use ethics consultation to ensure planning and delivery of the best quality and most appropriate patient care possible, in a manner that also attends to care-providers' wellbeing and the broader community's concern for the system.

Ethics consultation sometimes acts as a form of dispute resolution or as a lever for dealing with difficult circumstances of moral distress. Systematic, thorough and practical procedures of inquiry and deliberation that are deeply respectful of differing perspectives can be useful in situations of strained relationships and significant imbalances of power. With this combination of characteristics, ethics consultation is increasingly appreciated as a unique resource and an asset that can make a difference in large institutional settings.

Ethics consultation helps those involved to live with greater integrity. Facing difficult situations is part of the moral life of all human beings. This is true for members of care teams as much as it is for patients and families. Although participating in difficult health care situations is painful, those involved can develop integrity by working through some of life's most challenging questions. In this journey, care team members and patients and families need each other's partnership, for support, growth and understanding. Participation in an ethics consultation is not only an effort at problem solving; it can also be a transformational process for all involved. The evolution may be a change or broadening in perspective, or greater conviction regarding a previously held idea. The possibility for growth depends in part on the number and quality of encounters we have with people who hold different perspectives about what the context looks like (facts), and what should be most important (values).

1.2 Living with Integrity

For many years, my wife would laugh sympathetically when people asked me what I did for a living. That's because it would take me half an hour to go through the litany of activities I'm involved with as a bioethicist. Eventually I came to really understand my role. As a professional ethicist working in a practical setting, my job is to help individuals, teams, institutions and society more broadly to live with greater integrity.

People from all walks of life recognize and value the concept of integrity. In simple terms, integrity is the alignment of our beliefs and our actions. It is about living our beliefs in the decisions we make every day, big and small, personal and professional. In other words, living with integrity is about walking our talk in all aspects of our lives.[1]

This alignment is usually out of sync and rarely, if ever, perfect. This is because developing convictions comes with experience and evolves over time. As we go through life, we have new experiences, meet new people and find ourselves in new situations. Sometimes we realize that things we dismissed in the past have greater significance than we thought. Or we realize that new ideas matter more than do previous commitments. As time passes, our understanding of the world and our place in it broadens and deepens (hopefully!), and living with integrity looks different than it did before.

Integrity matters to everyone. We live and work with others and the choices we make separately and collectively impact the people who are part of our lives. This is clear in health care, where patients, families and loved ones, care providers, system administrators and even the public have to face difficult questions around clinical practice and policy. And they have to do so together.

Yet when we encounter situations where our integrity feels compromised, we tend to forget that others have important interests as well. We feel (often with intensity) *our* passion, *our* convictions, and *our* concerns. But the perspectives and feelings of others are not so easily accessible to us. Sometimes we don't even know who others in the situation are. However, *all* of the people affected by a situation have their integrity at stake.

Living with integrity is a lifelong struggle to deepen and refine our understanding of what life is about. Facing difficult situations is part of the moral life of all individuals. Thinking about ethics as the journey towards living with integrity recognizes the relevance of a situation to all of the people impacted by it. This approach to ethics creates room for people's stories to emerge in dialogue and be productively shared. In this way, people with different convictions can all find their place under the umbrella of integrity.

Living with integrity requires understanding how we see the world. To live my beliefs, I first need to understand the two broad types of belief I have and what my actual convictions are within these. The first broad type of belief concerns the way I see and understand the world. The second broad type of belief concerns values.

Descriptive beliefs about the way I see and understand the world range from big, abstract ideas to everyday understandings. They also span time and include our understanding of causal relationships between events. Beliefs at the more abstract

---

[1] See Cox, La Caze, and Levine's *Integrity and the Fragile Self* for an overview of philosophical conceptions of integrity (2003).

end have to do with what we think about the purpose of life, what happens when we die, whether or not there is a God, how we are connected to one another, and other such grand ideas. Such beliefs may seem philosophical and irrelevant to our everyday lives. But in fact, they have direct implications on the life choices we make.

For example, there is a range of possible beliefs about what happens when we die. Some believe that one comes back to earthly life in another form. Others believe that one transitions to an eternal, spiritual existence or some kind of communion with a higher being. Still others believe that one just disappears, full stop; one's body goes into the ground and that's the end of the story. Whatever our view, what we believe happens when we die will have an enormous impact on what we think a good death looks like, what a good dying process looks like, and how we think the last phase of life should be spent.

Sometimes in health care, participants in a patient's story will agree on the patient's diagnosis, the patient's experience of pain and suffering, and the likely outcomes of possible interventions. Yet, they will disagree about whether or not to use an intervention to pursue the continuation of life. This could be because they have different beliefs about what kind of quality of life is worth the patient's struggle and at what point in the quality-of-life spectrum the state of death is preferable to life. Assuming that they are able to put their own interests aside, those advocating for the intervention probably believe that the quality of the patient's life is better than death. Those advocating against the intervention believe that death is a better state of being than life of this quality. Notice that our assessment of the relative value of a given quality of life will depend on what we think happens when we die. So what looks like a disagreement about an intervention is in fact, at least partially, a disagreement about what happens when we die. In general, our opinions about what interventions we should employ in a health care context turn out to reflect beliefs about life's grand questions!

Examples of everyday beliefs about the world include things like where my car is parked, what the fastest route is from one place to another, and where I can find groceries at the lowest price. In health care, these include a patient's diagnosis and description of her symptoms. *I believe this woman has liver cancer and she is not experiencing much pain.* Basic beliefs also include descriptions of which health care providers and family members are involved with supporting the patient. *I believe the doctor looking after my mother is Dr. Hutchinson and I believe her clinic is open from 8am to 4pm.*

Causal beliefs concern the *consequences* of our actions. That is, we have beliefs about the causal relationships between sequential events. For example, I believe that in the early 1970s a strong nationalist movement prevailed in East Africa and led to my family moving to Canada in 1972. I believe that if I write this book, somebody will read it. And if somebody reads it, it will change his or her practice of ethics consultation.

Beliefs about causal relationships between events are central to health care. We are always intervening in people's lives, and our understanding of the cause of an illness or disease often impacts the interventions we choose. *I believe that this woman smoked a pack of cigarettes a day for 30 years and that is why she has lung cancer. I believe that surgery will remove the cancer, and that if she stops smoking it will help prevent the cancer from coming back.* For patients and providers alike, our attitudes toward interventions in health care are directly affected by beliefs about anticipated consequences. *I am open to chemotherapy because I believe this treatment will kill off the cancer cells in my body and without it I will die. I don't want surgery to remove the cancer in my prostate because I believe I may die during the surgery or it will leave me impotent.* It is easy to understand why sometimes there is disagreement about whether or not to use an intervention because some believe the intervention will have one consequence, while others believe the result will be different.

Thus, our beliefs about world range from the deep to the pedestrian and they are related together in time.

Living with integrity requires understanding our values. The second broad type of belief that guides us concerns values. There are many ways of talking about what should matter in life. We could talk about ethical theories, moral traditions, and the principles that these approaches highlight. We could also talk about moral rules–specific prescriptions for how we should act in given situations. We could talk about virtue or the character traits that good people should cultivate and demonstrate habitually. In my experience, when trying to solve practical problems involving people from diverse backgrounds and with a variety of perspectives about life, the most useful concept for thinking about and discussing what matters is the language of values, where a value is simply understood as something that is important.

To live with integrity, we must be able to understand the values we are demonstrating to the world. We also need to be able to critically examine what should matter in life. We need to be able to balance competing values we have and then demonstrate our considered values through our decisions, actions and attitudes.

Rather than naming values with just one or two words, like 'fairness' or 'respect', a good way to work through issues is to be specific about what matters in a situation. The more specific we can be, the better. For instance, imagine you are driving to get to an appointment. You are feeling stressed because you are very late. The pressure leads you to drive fast and just a little recklessly. Your response reflects what you understand the world to be like and what is important to you. (This response is likely not conscious–it is happening at a deeper level. But I'll get to that in a moment.) You begin to drive fast because it is important that you get to the appointment on time. Your peace of mind is disturbed because you fear that you may lose out on whatever is at stake in the appointment.

In the above example, relevant value statements might include 'it is important that I get to the meeting on time,' 'it is important that I don't keep people waiting for me,' and 'it is important that I don't cause an accident'. These specific statements can be thematized at a later stage under broader value headings (such as 'respect for others' and 'public safety') and defined in more general terms ('it is important to protect and advance the well-being of others').

Things matter for strategic reasons and/or for their own sake. Our values can be important to us in two ways: intrinsically or instrumentally. Something is intrinsically or inherently important to me when it is an end worthy of pursuit–when it is important for its own sake. Something is instrumentally or strategically important when it is a means towards ends of greater value to me.[2] For example, if you think that it's important for you to be kind to other people so that they will be kind to you in return, then kindness is instrumentally valuable to you. It's a way of getting what's really important to you, which in this case is to be treated well by other people. If, on the other hand, you think it is important to be kind to other people irrespective of the consequences, whether or not it leads to others treating you better, then kindness is more intrinsically or inherently valuable to you. In other words, it is valuable for its own sake, not because it gives you something else of greater importance.

One can begin to identify different levels of value commitments in the example of getting to the appointment on time by asking the question, 'why is this important?' It is important to get to the appointment on time, but why? It may be that the appointment is a job interview, so it is important to get employment. We can then ask again, why is this important? The answer will again depend on the context. Let's say that you are a single parent with a small income–getting the job matters because it is important to meet the basic needs of your family. And so on.

The ability to discern between intrinsic and inherent values and actually understanding the ends our more surface-level concerns are aimed at in our everyday choices are important for several deeper reasons. (Notice the instrumental/intrinsic distinction at play!)

Our beliefs about reality and about values are connected. What is important to me in almost any decision of consequence is tied up with my beliefs about the purpose of life and what it means to be a good person. In other words, where we turn for guidance about what should matter most will depend in part on our beliefs about the world. My beliefs about a deeper reality, including the existence of a higher being, a spiritual life, the nature of human responsibility, the connection between human beings, etc., will create the framework for my beliefs about what should matter in life.[3]

Someone trying to decide on what cancer treatment plan to consent to will have to weigh such considerations as minimizing pain and suffering, maximizing quantity of life, minimizing negative impacts on others, maximizing life experiences, and so forth. Each of these is a value. Each is significant and having to sacrifice any will be painful. In order to evaluate the relative merit of these considerations, I need to have an understanding of what it means for me to be a good person in general, and a good person with respect to the various roles that I play in life. For instance, how would a good father weigh the relative impact of maximizing his quantity of life, minimizing inconvenience and financial burden on his family, and being present

---

[2] See "Intrinsic Value and the Good Life" in William K. Frankena's *Ethics* for a discussion of different types of values.

[3] Charles Taylor discusses this relationship in his book *Sources of the Self: The Making of the Modern Identity* where he argues that an individual's identity is inextricably tied to the individual's understanding of the good.

for key family milestones? What it means for me to be a good father will depend on what meaning being a father has for my life. This is true for all of the roles I play and for my life in general. What it means for me to be a good person will depend on my understanding of the meaning of human life.

Human beings like to believe we can escape these big questions and just get on with things. But having a meaningful life comes from having a deep understanding of our place in the world. This yields an understanding of what should matter in life, which in turn enables making values-based choices. To paraphrase Socrates, it is examining life that makes it worth living.

Integrity requires articulating and challenging our beliefs. As I alluded to earlier, the journey towards living with integrity also happens at a subconscious and bodily level. We consciously create stories as we go, to crystalize the meaning of our experiences for ourselves and others. This haphazard dance between the conscious and subconscious, together with unsystematic reflection, leads most of us to operate with inconsistent beliefs.

Living with integrity is not just about walking any talk we might happen to have. Rather, it is about walking a talk that we have good reason to have. When it comes to our understanding of the world, believing something does not make it true. A belief that the world is flat or that the Toronto Maple Leafs won the Stanley Cup last year, no matter how passionately held and defended, does not make it a fact.

Often our beliefs about how the world looks are very hard to recognize, let alone analyze. Most of the time, these beliefs operate in the background and often develop without our even realizing it. Because these beliefs directly inform our preferred way of moving forward in a situation, and because we don't all share the same view of reality, not being able to recognize our own beliefs or commitments makes agreeing on the best course of action very difficult. Therefore, it is crucial that we figure out our beliefs about the world and that we are able to talk openly with others about them.

Exploring the justification of different types of beliefs about the world requires discussion and skill. Some of our beliefs about reality are good candidates for scientific confirmation. For example, the impact of a drug or an intervention can often be studied and tested in various ways. For this type of belief, the more evidence that we have for it, the more likely it is a *fact*. A useful discussion of the impact of healthcare treatments will require the skills to be able to identify relevant studies and experience and evaluate these to understand what they say about the world and how these facts relate to the issue at hand.

Some beliefs, on the other hand, are matters of *interpretation*. Beliefs about the purpose of life and what happens when we die, for example, are not good candidates for scientific analysis. They require a different sort of evaluation. Disagreements about these types of beliefs often cannot be resolved in a short space of time, if at all. While (thankfully) it is not necessary to resolve this type of disagreement for decision-making, understanding relevant others' perspectives about matters of interpretation is often crucial to building the trust that is necessary for collaboration and moving forward in complex situations. Empathy, deep listening and respectfully sharing one's own perspectives are important for making these conversations successful.

Integrity requires a pluralist response to diversity. Two people could arrive at the same situation and respond very differently. This is because in today's world we live and work with people who are different from us. What's more, we regularly make important decisions in contexts where others are involved. There are many ways we can respond to this difference, from insisting that others become more like us, to letting people be and insisting we should all go our separate ways.

The *Clinical Ethics Consultation* process, focused on integrity, is based on a pluralist response to diversity. This view insists diversity is a good thing, that we can learn from each other, and that we actually need each other to live with integrity.

Ethically justified decisions require the best available understanding of the facts and a well-considered understanding of what should matter most. Because of the limited perspective each person in a situation has, ethically justified decisions require those with a significant stake in an issue to identify, contribute and discuss their respective beliefs about what is *true* and what is *important*. Consciously crystallizing our own convictions requires reflecting on them set against the thoughts of others. We need to be able to focus on ourselves, listen to our hearts, look at the way we are living our lives, and think through what should matter most. We also need to partner with others because, in large part, we develop an understanding of what should matter through conversations with other people. And this needs to be done critically, thoughtfully, systematically and in a way that is attentive to our feelings.

Determining what should matter in life is a deeply personal experience and, at the same time, an inescapably relational activity. It calls for using respectful dialogue to go beyond dogmatic, timeless positions on issues to generate shared responses to shared challenges in actual contexts. In other words, an approach to ethics based on integrity makes room for difference. It requires finding agreement amongst people, even when their perspectives differ, in order to respond to difficult situations.

On the approach offered here, integrity also requires commitments to the values of trust, respect, humility, honesty, and courage. The *Clinical Ethics Consultation* process offers a way of working through ethical issues together in the midst of diversity that promises each of us the possibility of greater integrity.

To sum up so far, our beliefs are the stories we consciously develop that reflect our subconscious experience of life. They represent the way we perceive the world, what we think is true about life and the consequences we think various strategies of action will bring. We live according to these beliefs, altering them from time to time by hard-won experience. In order to live with integrity, we must be able to understand our beliefs and make sense of information that relates to our beliefs. We need to handle new information and change our views when presented with better evidence. And we need the ability to participate in conversations about these beliefs, big and small.

In almost all of the cases for which clinical ethics consultation is requested, participants have come to a new experience without a sufficiently well-developed understanding of their convictions. They are all trying to figure out what beliefs ought to guide the actions put into place to deal with the contentious issue. Conflict could be related to an issue or it could be related to how to work with others who have different perspectives on the issue. Participants are experiencing the misalignment of their existing convictions with their life experiences and are taking steps to try to shift towards alignment.

Integrity matters for peace of mind. Compromises to integrity lead to disturbances in one's wellbeing. Think of a ship whose structural integrity is breached. What happens? It begins to fall apart. It can't do what it is designed to do. Human beings are not so different. When what matters to us is enacted in our worlds, we feel whole. When values are threatened or lost in our lives, we feel torn apart.

Compromises to our integrity show up in our feelings. When I'm experiencing feelings of peace, contentment or joy, it's usually because what matters to me is showing up in my life. When I am scared, sad or angry, it's generally because something of great importance to me is in peril or lost. And when I'm confused and perplexed, it's often because I don't yet know what should matter to me. In this way, my state of mind and emotional wellbeing in any given moment can serve as a useful indicator of the extent to which what does matter to me is showing up in my life.

In situations where an ethics consult is required, participants are either struggling to figure out what should matter most in the situation, or they are feeling as though what matters to them is somehow in jeopardy. In either case, they are struggling to make meaning of the world and are likely feeling emotionally vulnerable (even if they don't show it).

We need integrity to achieve life goals. For those who want to follow a value system that includes achieving some end in life, paying attention to integrity is crucial. Beliefs about the purpose of life generate values that have to be understood and interpreted in order to achieve these goals. For example, those who believe the purpose of life is to become close to God will want to figure out how to achieve this goal as they face the vagaries of life. Indeed, most of what we commonly pursue in life has only instrumental importance. Financial wealth, power, and fame are not valuable in and of themselves but because of deeper benefits they provide. Those who agree that the benefit of these goods lies in something deeper will want to determine what the real value of these goods is, so as to appropriately frame their pursuit.

We need integrity to achieve the goals of our practice. In order for individuals on a health care team (or any team) to be able to figure out how to make decisions, such as how to allocate and use scarce resources, they must have a clear and justified understanding of their purpose and the deeper values they seek to live by. If this understanding is absent, then serious harms can and likely will occur at multiple levels.

One possible harm is that the kind of treatment offered to patients with similar conditions may be inconsistent if it depends on the particular values and beliefs of individual care providers in the system. Another harm is the moral distress felt by those with relatively less power on the team who must carry out decisions made by those with greater power with whom they disagree. A lack of clear and shared expectations and goals creates difficulty for all involved. Integrity will help us achieve our goals.

1.3 Key Assumptions of This Clinical Ethics Consultation Method

Several ley assumptions underpin the methods and goals of any clinical ethics consultation service. My proposed model for clinical ethics consultation assumes that:

- Knowledge is situated and comes from engagement with difference
- Deliberation grounded in trust and respect is necessary for integrity
- Ethics is the journey towards more ethically justified perspectives and solutions

These assumptions, discussed below, provide a foundation for the concept of ethical justification and the model for clinical ethics consultation proposed in this book.[4]

1.3.1 Knowledge Is Situated and Comes from Engagement with Difference

Every perspective is situated. How each of us sees any situation is limited by the subjective range of experiences that inform our perspective. There is no objective point from which one can determine the right decision. Each of us is embedded in a web of inter-dependent relationships. From the time we are born, we experience the world from within this web. With each encounter, we attach meanings to things, ideas and experiences, and gradually make sense of the world. In this way, our views are shaped by our own personal histories. While we may be able to recognize this limitation and think beyond our personal histories, the broadest vantage points we can achieve are still inescapably situated. In short, the meaning of any event or situation is limited by the history of the interpreter.

No individual has a purely objective view. This is true for health care providers, patients, loved ones and family members of patients. It is equally true for any 'ethics expert' brought in to provide support.

Knowledge is theoretical and experiential. As we become embedded in new situations, experiences impact our bodies, hearts and minds. Knowing our experiences first happens at a pre-conscious level. This knowing is inarticulate and at a visceral, bodily level. We then develop stories and our pre-conscious understandings become conscious. This alignment of the conscious and the pre-conscious is

---

[4]This approach aligns with hermeneutic ethics in the tradition of Hans-Georg Gadamer 1975 and Jürgen Habermas 1991. Guy Widdershoven reviews hermeneutic ethics in the clinical consultation context.

never complete, as we make sense of things more quickly than our cognitive faculties can process.

Consider the experience of being unwell. I often know I am unwell because of how I feel. But when I tell my wife or doctor and they ask me to explain my symptoms or describe my pain or discomfort, I can't always articulate these immediately in words. And when I begin my story, the words I use are often wrong or almost always do not completely capture the experience.

In any health care encounter, each of the individuals involved experiences the encounter pre-consciously at first. They then capture their experience in words. This happens as they communicate with others or for the formal record (e.g., the chart note). When the ethics consultant comes on board, she experiences the story as a new participant in the encounter, just as any other person in the story. Part of what is unique about the ethics consultant however, is that she consciously attends to the exercise of meaning-making.

Every solution is situated. So, it is though our experiences that we gain perspective. It is through the experiences we have that life takes on its meaning and that we come to develop understandings of the people we are with and the activities in which we are engaged. When a solution to a real life situation is developed, it is situated within the perspectives of all of those involved.

That knowledge is situated is also true of theories that emerge about how we should live. These ethical theories are created within cultural and linguistic contexts, at particular times and places in history, and are based on the situated interpretations of theorists. The norms of behaviour within a culture, what is understood as the way we should act, result from the interaction between reasoning and context.

So when someone (like a professional ethicist) uses an ethical theory to justify a solution to a challenging situation, it is not a 'clean' application of objective truth. Rather, the ethics expert is interpreting the theory and situation through the limited frame of reference from which that expert views the world. Any solution arrived at by an ethics expert is just as situated as any other solution.

Emotion and ambiguity are avenues for deeper understanding. While it's true that our responses to the world are informed by what we believe, we are not just intellectual creatures. We feel as we go through and respond to life's experiences. Our emotions–the visceral experiences we have–are connected to our beliefs about reality and to our values. Each of us has values–things that matter to us significantly. The extent to which we believe we have or will get that which is important to us determines our feelings. Our feelings are helpful indicators of our beliefs. When you see me feeling sad, you should know that there is a good chance I believe I've lost something important to me. You can ask if I am sad. And if I am, you can explore what I've lost that is causing me to feel sad. If I'm able to answer your questions, you'll be able to learn a little more about my values.

Our feelings impact the way we react to situations and express ourselves. Our feelings come out in our body language, facial expressions, and how and what we say. For example, how much I speak, the tone and volume of my voice and the speed and strength of my physical movement will likely change depending on what I am feeling and how strongly I am feeling it. When we are experiencing painful emotions we are at the highest risk of acting in ways that cause others pain. If I am feeling scared or sad, I may withdraw from others. This may cause fear, sadness or frustration for those who want or need us to partner with them on shared concerns. If I am feeling angry, I may express this with loud and hurtful words or gestures. This could lead to feelings of fear and anger in others as well, who might believe I am threatening their safety and wellbeing.

We need to be able to open our hearts to try to understand how people might be feeling, without judgment. We need to be aware of our own emotions and allow ourselves to feel them, also without judgment. For others and ourselves, we need to be able to name and acknowledge feelings. We need to be able to understand the beliefs that lead to these feelings. We need to be able to express and react to our feelings in ways that do not cause harm, but rather enable collaborative solutions.

People may not have considered the relationship between their values and emotional reactions. In most situations where an ethics consult is requested, the parties involved are in a space of un-clarity. They are enveloped in a fog of uncertainty and cannot see the right course of action. Understanding the most ethically justified way of proceeding requires entering this space and helping people to understand the contexts and values that should matter. Ethics consultants support participants in feeling and expressing pain in ways that avoid hurting others.

1.3.2 Deliberation Grounded in Trust and Respect Is Necessary to Live with Integrity

Growth in our perspectives requires collaborative relationships. Our values are not pre-formed and static. Rather, we enter a situation with *partially shaped perspectives* of what is going on and what is at stake, based on our experiences to that point. Then, during the course of interactions in the situation, our beliefs about reality and what is at stake develop further as we encounter new experiences and competing perspectives. Our understandings of facts and values are affected by our *relationships*; these understandings evolve as we interact with different people. We learn from each other. We hear what others think, process and evaluate their thoughts and further develop our own beliefs. And we respond emotionally as our understanding of our values deepens and our understanding of the alignment of the world with these values changes shape. In this way, exposure to others' critical and constructive perspectives enriches our own perspectives and leads to greater meaning.

Ethics is sometimes understood as the analytical activity of judging perspectives to be right or wrong. This approach is unhelpful according to the model of clinical ethics consultation described in this guidebook. Instead, a core function of the ethicist and the consult process is to help articulate accurate stories. This is because dealing with difference productively requires helping people make meaning. It requires helping people to come together to develop shared understandings. This requires great sensitivity and openness to different perspectives.

To illustrate this point, I'll share a personal example of advance care planning. In considering my own life and what will be important to me during my last days, I realized that what gives me great joy in my life is thinking about complicated ideas and then trying to communicate them to others. I like to think, I like to help others think more clearly, and I enjoy helping people solve difficult problems. This is in part because it's very important to me to make a difference in the world. I also believe things are pretty good on the other side of death. So I told my wife that should I lose my cognitive function through illness or injury, please don't try to extend my life–just let me die. I was feeling quite confident and proud of this work I had done…until she pointed out that my attitude was rather selfish. She said that in such a situation I may not be able think my high falutin' thoughts, but she and my son may still be able to take care of me, to love me, and to enjoy something of a relationship with me. Faced with this new idea, I realized that while thinking, working, and helping are important to me, most important in my life is my relationship with my family, and their joy and satisfaction in life. So, of course, my request changed. This dialogue was crucial to helping me figure out what matters in life.

Our relationships with others can and do change us. We enter into an encounter with one view of the world, share experiences and exchange ideas, and leave differently than we came. The difference can be a change of perspective. It can also be a deeper commitment to ideals previously held. I know for me it is just these types of conversations that help me to make sense of my life over time.

The process of clinical ethics consultation described in this guidebook is intended as a transformative intervention in the lives of the participants. The method seeks to help participants recognise that their views are partial and incomplete. It seeks to build participants' trust that their perspectives will be seen as worthy of engagement and to create a safe space where they will be heard and given room to reflect, think and grow.

Collaborative Relationships Require Trust and Respect. If people are scared that what matters most to them–their wellbeing, their integrity–is not of interest to others, they will be disinclined to partner with these others. They will adopt a defensive posture to shield possible vulnerabilities. Of course, this prevents their growth, but to the end of self-preservation. *Trust is the value that speaks to our comfort with being in community.* When people trust that what matters to them will be taken seriously, when they believe they can count on others to help them make meaning of difficult circumstances, they are more inclined to be open to partnership.

The ethics consultant must help nurture relationships of trust in order to help people come together to work through the challenging situation they are in and arrive at the most ethically justified decision possible. The key to building trust is engendering relations where people treat each other with respect.

In doing this, the ethics consultant is up against some difficult barriers. First, this important value–trust–is exactly what is often compromised in situations where ethics consults are requested. Not only has some sort of communication or relationship breakdown most likely taken place, but, in my experience, often patients' and families' mistrust is a result of previous negative experiences with the health system.

The second barrier concerns power dynamics. In almost all health care encounters, health care team members have expertise about the technical dimensions of the situation or knowledge about the health care system. Patients also have key expertise–about their own life stories and the meaning of their health issues within those stories. All those involved have some understanding of what a meaningful life looks like and what values we should live by. And integrity is at stake for all of them; they deserve to be engaged with and heard. But the health care system has a hierarchical culture that empowers the voices of some over others in ways that do not necessarily correspond either to technical expertise or relevant values.

In the health care context, patients are vulnerable to care providers in at least three ways: (a) patients are likely unfamiliar and uncomfortable in the surroundings where the situation is taking place–surroundings care providers spend every day in; (b) the wellbeing of patients is usually compromised, while care providers are usually healthy; and (c) patients have to rely on health care providers' expertise, whereas health care providers enter into the relationship precisely because of the skills and knowledge they possess. Because of this power differential, health care professionals have a greater responsibility to engage carefully and manage this power imbalance as much as possible. Patients are owed a fiduciary duty by professional caregivers. Being a professional in this context means accepting an obligation to protect patients' trust and wellbeing.

Ethics consult team members must manage power dynamics in these difficult situations to build trusting relationships so participants can engage in systematic analysis. Building trust is best done by helping the people involved to treat each other with respect. In my view treating others with respect requires three kinds of behaviour. The first is being kind and caring. This means treating others gently and well, regardless of whether or not you agree with their words or deeds; it means treating others as if they had as much or more power in the relationship than you (even if they don't). Kindness and caring does not mean that there is no room for anger, frustration, or other strong emotional responses; rather it requires that these be conveyed within a kind and caring relational framework.

Respect also requires listening without judgement, putting aside your convictions, no matter how strongly you feel them, and opening your heart and mind to try to understand and feel the other's perspective. Also known as empathetic understanding, this second requirement of respect is about listening to others, seeking first to understand his or her perspective before trying to settle on values or solutions.

The third behaviour respect requires is collaborative engagement. This means showing up and sharing your beliefs and the reasons for them, and seeking to work together with others to develop a broader perspective from which to act. To respect others is to take seriously their ability and interest in making justified decisions by sharing competing perspectives and pushing them to take these alternate views seriously in the development of their own perspectives.

In contexts where we have much greater power over others, engagement can easily slip into coercion, where others are forced to change their perspectives not because they are really convinced, but because they are afraid of being harmed, or at least not helped. So this dimension of respect has to be applied with great care in contexts where there are significant power imbalances between the parties–as in therapeutic relationships.

Central to this clinical ethics consultation method is treating people with respect in part to build trust so they can participate in a collaborative exercise to solve an issue knowing they won't be forced to change their view or have that view dismissed without being engaged.

1.3.3 Ethics Is the Journey Towards More Ethically Justified Perspectives and Solutions

Clinical ethics consultation should see ethical problems as "inevitably situated and structured by the implications of interpersonal interactions, the semantic, institutional and political context, socially structured meanings and pre-understandings" (Ohnsorge and Widdershoven 2011). Given this reality, ethics should not be seen as the theoretically driven enterprise of defining the right answer to a practical problem according to some external standard. What ethics can achieve, through the service of clinical ethics consultation, is helping people negotiate ways of moving through ethical problems based on evolved, shared understandings of the situation set against the broader moral norms of society.

Clinical ethics consultation leads to shared understandings. One might see ethics as the process of making the negotiation between our subconscious experiences and the conscious development of our stories about the world explicit, intentional and in partnership with others. Clinical ethics consultation serves this end, helping participants involved in a case operate with a clear and shared understanding of their respective beliefs about the way the world looks. The process enables shared interpretations of evidence where possible, and the identification of divergent interpretive understandings where common ground cannot be established.

Ethics consultation should aim to facilitate dialogue between people involved in an ethically complex situation. This dialogue should help them to develop a common moral horizon and to build a solution that brings them closer to this horizon. The ethics process should help people establish shared commitments and problem-solve together.

This guidebook proposes a clinical ethics consultation process that leads to solutions grounded in context. The process merges subjective, individual standpoints into collectively shared understandings of the facts and of relevant values, informed by social moral norms. The process creates room for dissenting voices and space for participants to deliberate–to hear each other, to argue, and to reconstruct their own views in response to these exchanges. Ethics consultants using this process take participants' stories seriously and resist making assumptions. They also introduce moral and legal frameworks into deliberations, and seek responses to these when negotiating a way forward. Further, the process holds that solutions are not predetermined and can be creative and innovative. While the ethics process outlined in this book aims to help people work together to make their own clear and deep meanings out of difficult situations, these solutions are resistant to concerns about relativism. This is because the process itself is based on the foundational values of integrity, trust and respect within a framework of accepted social and moral norms.

Patients, loved ones, and care team members are moral agents. This negotiation of meaning and shared decision-making leaves accountability for decisions in the hands of participants in the process. Making an expert accountable for decision-making neglects the moral agency of others in the situation. Moral subjects are able to make sense of a situation, evaluate what matters, and make decisions accordingly. In almost all difficult cases, patients, loved ones and care providers are the moral subjects impacted by decisions. Indeed, in situations challenging enough to warrant ethics support, their moral agency is on the line to a significantly greater extent than for the expert resources brought in to provide support. This is because their talk is going to be shaped through the experience.

Sometimes in thorny situations people don't want the responsibility of making difficult or painful decisions. This is understandable, and people can reasonably choose to forego this responsibility. But adopting a method that takes the possibility of decision-making out of their hands is not acceptable. Simply removing moral responsibility from the people involved in a situation disrespects their moral agency and their dignity.

Taking decision-making out of the hands of those involved enables the abdication of moral responsibility, risks ineffective solutions where an enforcement mechanism is not in place, and, where the solution can be enforced, increases the likelihood of moral distress. People are more likely to comply with decisions they have helped to make. People experience moral distress–feelings of anger, powerlessness, guilt, and frustration as their integrity is compromised–when the decisions of other are forced upon them.

1.4 The Ethics Consultant

Following from these starting points, the model of ethics consultation offered in this guidebook sees the function of the consultant as an expert of both ethics and facilitation. Widdershoven et al. (2009) suggest that the roles of ethics consultants include interpreter, educator, facilitator and Socratic guide.

In the interpreter role, the consultant helps assign meaning to the issues at hand. He or she works within and between the perspectives of those involved and broader understandings of ethical concepts to help establish what is actually happening and at stake in the situation. The facilitation role, in my view the most crucial, involves establishing the terms of engagement with participants in the consult, managing power differences, deeply listening in one-on-one conversations and then bringing people together to hear each other and collaboratively move through the difficult questions and issues being faced.

1.4 The Ethics Consultant

The role of educator involves gathering information on a variety of subjects, from clinical and ethical concepts, to the perspectives of those involved. And as a Socratic guide, the consultant slows things down and probes participants' perspectives and assumptions. These roles are sometimes challenging to balance and require skill and sophistication. It is also important to remember that consultants themselves grow and change through their involvement in each consultation process.

As I suggested earlier, ethics is a way of looking at life. Ethics consultants, in the brief encounters they have with those involved in consult situations, help participants to understand this idea. Consultants use skill and wisdom to help merge a disparate set of stories into harmony for a moment. They create space for respectful dialogue and collaborative deliberation to help others resolve problems in a way that enables all participants to live with greater *integrity*. To do this effectively, ethics consultants need to master four types of competencies:

1. Listening and understanding
   - The ability to listen to and understand the stories, values and emotions of others, as well as what they believe are the facts.
   - The ability to understand their own stories, the facts and values as they see them, and their feelings.
   - The ability to acknowledge and support their own and others' emotional experiences.

2. Reflection and assessment
   - The ability to critically reflect on what is true about the world, based on evidence, interpretation and reasoning.
   - The ability to critically reflect on what is important in life from multiple theoretical and practical perspectives, and to balance different values.

3. Intervention and support
   - The ability to facilitate discussion to help others get a clearer and deeper understanding of their own beliefs.
   - The ability to bring people together through discussion to understand each other's perspectives, in contexts of difference and inequality.
   - The ability to enable people to identify and build common ground.
   - The ability to enable people to build solutions to shared problems.

4. Action and articulation
   - The ability to make and implement decisions based on well-considered facts and values, even when doing so is difficult.
   - The ability to articulate perspectives, choices and rationale (the justification for choices).

The American Society for Bioethics and Humanities' *Core Competencies for Health Care Ethics Consultation* provides a more detailed list of core competencies necessary for health care ethics consultation. It sets out the core skills, knowledge and character traits that ethics consultants require.[5]

In this resource, skills are divided into:

- Ethical assessment skills
- Process skills
- Interpersonal skills

Core knowledge falls under categories of:

- Moral reasoning and ethical theory
- Common bioethical issues and concepts
- Health care systems
- Clinical contexts
- The local health care institution
- The local health care institution's policies
- Beliefs and perspectives of local patient and staff population
- Relevant codes of ethics and professional conduct and guidelines of accrediting organizations
- Relevant health law

And the following character traits important for effective ethics consultation are set out as:

- Tolerance, patience and compassion
- Honesty, forthrightness and self-knowledge
- Courage
- Prudence and humility
- Integrity

Ethics consultants should be aware of this guiding resource. Before leaving this chapter on the ethics consultant, it is crucial to clarify for those engaging this work that reading this text or following the toolkit is not sufficient to do the work well. If anything, this guidebook should underscore the point that clinical ethics consultation requires skills and knowledge in a variety of areas. It also requires time and resources. Not having the training or resources to do this work well risks causing significant harm–from reinforcing inappropriate power dynamics to undermining the integrity and wellbeing of those involved, including the family and institutions within which the situation is unfolding.

Ethics consultants must keep the consult role distinct from other professional roles. Ethics consult team members are not there to give legal advice, medical advice, political advice, or social work advice. Their task is to help people understand the values and beliefs that lead to tension and disputes, and to help them work towards a resolution. Increasingly, people from different professional backgrounds and perspectives have been invited to participate on ethics committees to

---

[5] A precursor of this list is provided by Francoise Baylis in "A Profile of the Health Care Ethics Consultant" found in her book *The Health Care Ethics Consultant*.

enrich the breadth of perspectives in discussion. Accordingly, many people who become involved in ethics consultation, though highly trained professionally, have little preparation for this new task. Such individuals face the challenge of wearing two hats: their professional hat, and their ethics consultant hat. Such team members must consider carefully what changes this may imply for the beliefs and values they accept and thus for the positions they might take within an ethics consult environment.

For example, a physician colleague was involved in a consult where a dying patient was not receiving sufficient pain medication to control the chronic pain he was experiencing. This hindered decision-making because the pain was impairing the patient's ability to understand his clinical situation, reflect on his goals for a good death, and decide on the preferred alternative for his end-of-life care. As a physician, my colleague very much wanted to pull out his prescription pad and order the pain medication that would improve not only the patient's quality of life but also his ability to participate in the consult decision-making process. But of course that is not the role of an ethics consult team member.

So the physician, acknowledging this personal tension to other team members, continued to pursue the steps of inquiry of the ethics consult. Afterwards, when he was not wearing the ethics consult team member hat, he contacted this patient's attending physician and indicated an appropriate prescription for pain medication. The family physician asked my colleague to write the prescription on the spot. This was an appropriate solution to the problem, as the member sought to keep the consult role distinct from other professional roles.

Not only is it important that ethics consultants should separate their ethics consultant roles from other professional roles they may occupy, but they would also be well-advised to critically reflect on how the assumptions and practices of their other roles align or are in tension with the assumptions on which the clinical ethics consultation model are based. In other words, as identified much earlier on in the text, all professional practices privilege certain values. Someone coming to the work of clinical ethics consultation is effectively merging two different practices. If the conceptual frameworks of these practices are not critically reflected upon, it could leave the consultant in moral distress. Put another way, the integrity of the clinical ethicist is very much on the line in this work as well.

Every ethics consult team member shares responsibility for creating space for respectful dialogue. This is not always easy to do. Today's health care context is busy and complex–not an easy environment for deep deliberation. People do not have the time or support to ask hard questions, grapple with uncertainty, or even just sit down, clear their minds, and listen carefully. That is why so much of the ethics consultant's work is simply to provide that conversational space.

As Margaret Urban Walker stresses in her well-known book *Moral Understanding: A Feminist Study in Ethics* (1997), it is really about keeping moral spaces open. It is often about putting on the brakes in a situation, recognizing that some very difficult and challenging issues must be faced collectively, providing a safe environment to have these conversations, and generally facilitating the deliberation process. In short, *an ethics consultant is needed to create within the health care system the space and opportunity to have the kind of conversations that in an ideal world could happen every day.*

1.5 Types of Consult Requests

All along the continuum of care, there are situations that would benefit from consult support. In such cases, consultants could help participants arrive at the best possible decisions–decisions that are ethically justified and allow all participants to live with integrity. Here are some examples of situations where ethics consultation could help:

- In acute care, a woman, age 70, has just had her third stroke and is now significantly demented and on a ventilator. Her three daughters disagree about the proper form of care. The physicians prefer one thing, other caregivers prefer another, and they all seek to reach a decision that can be defended as the best one for this patient. Ethics consultation could help determine the appropriate goals of care for someone who is seriously ill.
- In residential care, a relatively young man, age 50, is in long-term care for advanced multiple sclerosis. He is not able to swallow very well as a result of his illness and is at risk of aspirating–yet he wants to eat by mouth. The team is struggling with how to honor his wishes, support the family, and also deal with their own concerns about being the immediate agents of this man's death. Ethics consultation could help determine if living at risk is appropriate in this situation.
- In a community, a quadriplegic man, age 28, has been receiving home care for 5 h a week for the last six months. One day, in the process of doing basic housekeeping, staff members happen to find some illegal drug paraphernalia. They are unsure whom, if anyone, to share this information with, and whether this discovery should affect the kind, quantity, and quality of care that they provide. Ethics consultation could help determine how to best provide safe, quality care for all involved.
- In an administrative context, a medical director of a small community hospital well understands concerns about the approach of an influenza pandemic. She needs to discuss with her medical staff the need to work in a context of an increased demand for service and a shortage of service providers. She is not sure how to approach her colleagues about this and how to address the issue of work versus family commitments in times of a health crisis. Ethics consultation could help determine the best communication plan during a pandemic.

These are only a few examples of issues faced in various health settings. There is no shortage of ethically challenging situations in contemporary health care, and this is true in every quarter and branch of the system.

1.6 Types of Consult Support

Ethics consultation is still relatively unknown in many areas of health care. Even when people request an ethics consult, they may not know what they need or what to expect. For this reason, it is important for ethics consultants to have a clear understanding of what services they can provide, and the different ways in which these supports can be delivered. This chapter describes several basic types of ethics consultation support (Table 1.1).

The simplest, and often the initial, type of consult is the chance encounter or telephone call. A colleague approaches the consultant in the hallway or perhaps telephones the office and says, "I'm facing this issue and I'd like to have a quick chat about it, if you don't mind." This support might require anywhere between 10 and 90 min, often depending on the level of anxiety the caller is experiencing.

During the conversation, you discuss the issue, maybe sharing insights as you try to understand your interlocutor's perspective and deepen his/her own understanding. Your objective in this kind of situation is really to empathize with the person, provide them with a sounding board, and share information about what formal systematic consultations involve and what a decision framework looks like. You might also offer briefly considered and preliminary opinions about what relevant issues and values in the situation may be, or what conversations might be helpful for addressing tensions. The consultant's deliverable here is a conversation.

Table 1.1 Consult types

Type of consult	Meeting length	Participants	Objectives	Ethics deliverables
Hallway/telephone chat	10–90 min	Individual	Sounding board/briefly considered opinion	Conversation
Team – prospective	1–2 h	Team	Support towards decision	Facilitation through a process; list of values and implications
Clinical case	1–6 h	Team, Patient, Family	Support towards decision	Facilitation through a process; list of values and implications; detailed consult report with recommendations
Team – retrospective	1–2 h	Team	Support understanding of past decision	Facilitation through a process; list of values and implications
System/policy	1–6 h	Team	Support towards decision	Facilitation through a process; list of values and implications; summary of process and implications

The hallway chat or phone call is not a formal ethics consult, but it is absolutely appropriate. In fact, responding to such requests is essential for building the credibility and relationships key to the success of any ethics consultation service. However, these casual meetings are limited and lack the rigor required for ethical justification. Having contact with only one participant in the situation, the consultant can comment only on a very small piece of the picture.

It is important not to treat or refer to this type of meeting or telephone call as a formal consult. Such a misinterpretation could do significant harm. An ethics consultant once told me about a short telephone conversation she'd had with a physician about a relatively difficult case. The ethicist expressed some concerns around the handling of it and recommended to the physician that a formal consult be initiated. That never occurred.

Two months later, the patient involved went to another hospital where an ethics consult was requested. The patient's chart made mention of a previous ethics consultation and when the consultant was contacted it become clear there had been only a telephone chat. The first physician, apparently believing that ethics requirements had been fulfilled, had recorded the chat as a formal consultation and implied that the course of action at the initial institution had had been approved by the ethics service.

So it is crucial to be very clear about the limits of this type of intervention. A hallway or telephone conversation does not in any way substitute for a formal consult involving relevant participants and a full inquiry.

A team prospective ethics consult is requested when a care team anticipates that they are going to be facing some very difficult questions about the care of a patient. Perhaps they foresee some tension among or between family and team members. Or they may know that the patient's ability to participate in decision-making is going to be questionable and they want to think carefully beforehand about how to deal with decisions. The team asks for support in advance of facing these challenging issues.

In this case, the intervention might be a 1- or 2-h conversation with the ethics team, using a systematic process to help the care team consider what is most important in caring for this patient and their family, and discussing some possible strategies for living out these values. The parties involved in this kind of consult are usually limited to members of the ethics team and care team.

Patients are usually not participants in the team prospective consult, which is about the team coming together to get their bearing on the situation; this may be very difficult if the family and patient are present. However, the ethics consultants will remind the team any actual decisions about patient care will have to include the patient's perspective. In other words, while it may be appropriate for the team to work together to anticipate challenges to patient care, *the actual decision-making should be inclusive of the patient, surrogate decision-makers, and loved ones, as appropriate.*

The formal clinical case consult is a full-fledged consultation, involving all parties, where critical decisions will be made. There may be disagreement, tension, uncertainty, or conflict among participants, and the most ethically justified course for moving ahead may not be apparent at first. Taking a minimum of a few hours, and usually taking place over the course of days, this type of consult is extensive.

The clinical case consult usually includes the patient, family members or loved ones, as appropriate, and members of the health care team. On some occasions, outside support people including members of administration, risk management, and so forth may be involved. The deliverable in this kind of consult is the facilitation of a systematic inquiry into the beliefs of participants.

There are three dimensions to this work: descriptive, prescriptive and problem-solving.

Ethics consultation begins with descriptive work. We start by listening and trying to understand the perspectives of various people involved in the situation. This listening usually happens through a process of reviewing charts and interviewing participants. The idea here is to try to describe participants' beliefs about the situation.

The next stage is the prescriptive work of coming to a shared understanding of the facts in the case and the values any solution should live up to. Descriptive work surfaces beliefs; prescriptive work explores the beliefs that should guide decision-making, and whether shared beliefs are possible.

We want to share an understanding of clinical information about the situation. In particular, we want to make sure those without technical training and expertise are given sufficient information and support to be able to make sense of available evidence. This includes patients and families in the clinical context and possibly administrators and others in the system-level context. It may also include members of the ethics consult team. Once we have heard the perspectives of those involved, we can take time to identify any mistaken understandings, and agreements or disagreements about the facts. It is important that we name where people stand on quality of life. However, based on respect for autonomy and diversity, we do not need to agree about this. We simply need to know where people are located against the landscape of these broad beliefs.

The problem-solving stage considers how to bring behaviours and attitudes in line with the values participants want applied to decisions. This is where the group comes together and creates a solution that will allow everyone to live with as much *integrity* as possible. This phase needs to follow the prescriptive dimensions of the conversation, because absent a shared understanding of the standards a solution must meet, it's not possible to know whether we've arrived at a good answer. This step usually comes towards the end of the Consult Meeting stage.

The full-fledged consult will usually require all elements of the step-by-step process described below in Chap. 2. In a full consultation like this, a formal report would be completed to describe the process undertaken, the parties involved, and the decisions and recommendations made.

A team retrospective consult occurs when a team is uncomfortable with how things happened in a given situation and wishes to debrief the experience from an ethics perspective. Some difficult issues and questions arose over the course of a patient's care. These issues were addressed and the questions answered, but team members were divided about whether the right thing had been done, and some were upset. Members of the health care team thus asked for support from the ethics team to help clarify what exactly happened and how things might be improved next time.

This kind of meeting can take between 1 and 2 h, and involves the ethics and health care teams. Typically, the family is not involved because they have gone through the situation and no decisions to be made require their input. Indeed, because ongoing participation may cause them pain, it is usually best to leave them out of the process. Here again, the consultation deliverable would be facilitation' using a systematic process aimed at exploring team members' concerns. Ethics consultants may discuss the team members' feelings in terms of 'moral distress'. Again, moral distress is very much tied to our decisions or actions being in conflict with our deeply held values.

The consult process, in this case, seeks to help participants understand the values they could not honour in their care decisions. The process will also consider what other values were honoured in the decisions made and whether, on critical reflection, values were prioritized appropriately or not.

System level ethics consultations address issues that go beyond decisions for an individual case. Such consults usually deliver products such as policy, practice guidelines or team strategies. System-level issues can be organizational issues such as assisting institutions to determine criteria and processes for allocating scarce resources. They can be clinical practice questions related to establishing standards and processes for responding to challenging clinical issues. Or they can be team dynamic issues that require strategies for assisting programs to determine how best to work together.

Typically, in these cases, the team analyzing the issues recognizes some important ethics dimensions and asks for ethics consultation assistance. These types of consults usually take at least two, 2-h meetings, depending on the issue. Most often, only the ethics team and policy team are involved, unless wider consultations are required.

The overall process for administrative consults is similar to that for clinical consults. Consults for administrative issues are often (but not always!) less emotionally charged. The process involves the following stages: establishing the decision team, ethics analysis to develop a preliminary solution, engagement of others in the analysis process, articulation and communication of the decision and rationale, and implementation of and follow-up on the final decision.

A system-level consult is designed to assist leaders achieve trust, quality, efficacy, legitimacy, and compliance.[6]

---

[6] For an overview of the ethics dimensions of system-level issues, please see Bashir Jiwani's 2015 article on ethically justified decisions "Reflections on HealthCare Leadership Ethics: Ethically Justified Decisions. Healthcare Management Forum" 28(2): 86–89 and "A Further Landscape: Ethics in Health Care Organizations and Health/Health Care Policy" found in Storch et al.'s *Towards a Moral Horizon*. Bashir Jiwani's system-level decision-making tool, Good Decisions, is available at incorporatingethics.ca.

1.7 Models of Ethics Consultation Services

There are different models of ethics consultation, including the standing ethics committee, the clinical ethics consultant, and the multi-member ethics consult service. The process described in Chap. 2 is best suited to the practice of clinical ethics consultation by multi-membered consult teams. However, the process can also be applied in alternate models.

The *standing ethics committee* is fully involved in every ethics consultation alongside ethics consult staff. This model naturally depends on a very active and supportive ethics committee. When an issue arises, a consult is requested, and members of the consult team (independent of the ethics committee) do some preliminary work. Then everybody involved–patients, loved ones, and staff–is asked to come to a meeting with the ethics committee and ethics consult team members, where the issues are discussed thoroughly and thoughtfully. The ethics staff help the committee create a hospitable climate for the meeting.

This comprehensive approach brings together a wide variety of perspectives. However, this approach can impose heavy burdens on participants. For example, the patient and family face the additional stress, and perhaps humiliation, of having to tell their story yet again to another group of strangers. And there are significant logistical and resource costs associated with having a full committee of sometimes 12–20 people meeting together and deliberating in enough depth and specificity to provide real support on ethics consultations.

The model described in Chap. 2 can assist with the pre-work done by the ethics consultants in this model in preparing for the broad consult meeting and can also provide direction for what the larger meeting might look like.

The clinical ethics consultant **model involves an ethics-trained individual providing independent and direct support in multiple and various clinical contexts.** This formally trained ethics professional is available to provide direct support and can respond quickly to requests. The ethicist can provide leadership in the integration of values-based decision-making in the institution. Others in the institution who are involved in ethics consults learn how to handle situations effectively based on the support and expertise provided by the ethics consultant.

Yet there is an important disadvantage with this approach as well. It may appear to suggest that ethics, rather than being everybody's business, are only the concern of specially trained ethics consultants–a reversion to the segmented model of administration. This model may promote the idea that if a difficult ethical issue arises, an expert sitting in an office somewhere should be called upon to sort things out.

Almost all aspects the approach set out in Chap. 2 below can be applied in full by a clinical ethicist.

My preferred model for ethics consultation support is *the multi-member ethics consult service*. Here a team of individuals is charged with the responsibility of providing a first line of support to individuals and teams facing ethically challenging situations. The ethics consult team leads by gathering information about the situation, making a systematic inquiry of the relevant issues, and initiating a formal decision-making process.

The individuals providing this service have formal training in providing ethics support. They have listening skills and can enable respectful dialogue between different people. They have an understanding of different ethics perspectives and various values commonly at stake in these situations. These individuals will have developed the character traits appropriate for this kind of work. Further, ethics consult team members are able to employ a systematic, values-based decision process, such as the one described in Chap. 2 of this guidebook.

An important concern with this consult-team approach is that if team members are not sufficiently trained, the quality of ethics consultation provided will be poor– and might even do more harm than good. Also, if consult team members are unequally trained and skilled, the more competent members will carry an increasing proportion of the load, and become increasingly depended upon, which can lead to burnout. A small, multi-member ethics consult service should be supported by a professionally trained ethics consultant who can provide guidance and training on an ongoing basis, and also be called on directly in particularly challenging cases.

This layered approach has greater agility and responsiveness than the comprehensive standing committee approach, and a greater diversity of perspectives than the clinical ethics consultant approach. This model also encourages capacity-building, because members of the consult team are also members of the wider care giving community.

All three of these models can work well, depending on the circumstances. However, having worked with numerous ethics committees in Canada, operating from institutional, to regional, provincial and national levels, I have become an advocate of the *multi-member ethics consult service*.

Ideally, the multi-member team is sensitive to the local culture in the community and credible to community members. Team members can bring both experience in particular areas and outsider perspectives to the discussion of an issue. They can be adequately trained on issues, values, and theoretical dimensions, as well as in the skills of clinical ethics consultation. Then they can be backed up by a professional ethics consultant for training, advice, and, when necessary, direct intervention.

References

Andre, Judith. 1997. Goals of Ethics Consultations: Toward Clarity, Utility, and Fidelity. *Journal of Clinical Ethics* 8 (2): 193–198.

Baylis, Francoise. 1994. *The Health Care Ethics Consultant*. Totowa: Humana Press.

Beauchamp, Tom L., and James F. Childress. 2013. *Principles of Biomedical Ethics*. 7th ed. New York: Oxford University Press.

Cox, Damian, Marguerite La Caze, and Michael P. Levine. 2003. *Integrity and the Fragile Self*. Aldershot: Ashgate Publishing Limited.

Daniels, Norman. 1979. Wide Reflective Equilibrium and Theory Acceptance in Ethics. *The Journal of Philosophy* 76 (5): 256–282.

References

Fox, Ellen, Kenneth A. Berkowitz, Barbara L. Chanko, and Tia Powell. 2006. *Ethics Consultation: Responding to Ethics Questions in Health Care*. National Center for Ethics in Health Care. http://www.ethics.va.gov/ECprimer.pdf. Accessed 29 Feb 2016.

Frankena, William Klaas. 1973. *Ethics*. 2nd ed. New York: Pearson.

Frolic, Andrea, and the Practicing Healthcare Ethicists Exploring Professionalization (PHEEP) Steering Committee. 2012. Grassroots Origins, National Engagement: Exploring the Professionalization of Practicing Healthcare Ethicists in Canada. *HEC Forum* 24 (3): 153–164.

Gadamer, Hans-Georg. 1975. *Truth and Method*. 2nd ed. London: Bloomsbury.

Goodin, Robert. 1995. *Utilitarianism as a Public Philosophy*. New York: Cambridge University Press.

Habermas, Jürgen. 1991. *The Structural Transformation of the Public Sphere*. Cambridge: MIT Press.

Harman, Gilbert. 2000. *Explaining Value and Other Essays in Moral Philosophy*. New York: Oxford University Press.

Jamieson, Dale. 1991. Method and Moral Theory. In *A Companion to Ethics*, ed. Peter Singer, 476–487. Oxford: Blackwell.

Jiwani, Bashir. 2012. Good Decisions: A Map to the Best System-Level Decision, All Things Considered. http://www.incorporatingethics.ca/view-good-decisions.php. Accessed 23 Feb 2016.

———. 2015. Reflections on HealthCare Leadership Ethics: Ethically Justified Decisions. *Healthcare Management Forum* 28 (2): 86–89.

La Puma, John, and David Schiedermayer. 1994. *Ethics Consultation: A Practical Guide*. Boston: Jones & Bartlett Publishers.

MacIntyre, Alasdair. 2007. *After Virtue: A Study in Moral Theory*. 3rd ed. Indiana: University of Notre Dame Press.

Martin, D., and Peter Singer. 2003. A Strategy to Improve Priority Setting in Health Care Institutions. *Health Care Analysis* 11 (1): 59–68.

Ohnsorge, Kathrin, and Guy Widdershoven. 2011. Monological Versus Dialogical Consciousness – Two Epistemological Views on the Use of Theory in Clinical Ethical Practice. *Bioethics* 25 (7): 361–369.

O'Neill, Onora. 1975. *Acting on Principle: An Essay on Kantian Ethics*. New York: Columbia University Press.

Rawls, John. 1971. *A Theory of Justice*. Cambridge: Harvard University Press.

Rodney, Patricia. 2012. Moral Agency: Relational Connections and Trust. In *Toward a Moral Horizon*, ed. Janet Storch, Patricia Rodney, and Rosalie Starzomski, 2nd ed., 153–177. Toronto: Pearson Education Canada.

Rodney, Patricia, MaryLou Harrigan, Bashir Jiwani, Michael Burgess, and J. Craig Phillips. 2012. A Further Landscape: Ethics in Health Care Organizations and Health/Health Care Policy. In *Toward a Moral Horizon*, ed. Janet Storch, Patricia Rodney, and Rosalie Starzomski, 2nd ed., 358–383. Toronto: Pearson Education Canada.

Shah-Khazemi, Reza. 2007. *Justice and Remembrance: Introducing the Spirituality of Imam Ali*. London: I.B Tauris and Company.

Strong, Carson. 2010. Theoretical and Practical Problems with Wide Reflective Equilibrium in Bioethics. *Theoretical Medicine and Bioethics* 31 (2): 123–140.

Taylor, Charles. 1991. *The Ethics of Authenticity*. Boston: Harvard University Press.

Walker, Margaret Urban. 1997. *Moral Understanding: A Feminist Study in Ethics*. New York: Oxford University Press.

———. 2008. Groningen Naturalism in Bioethics. In *Naturalized Bioethics: Toward Responsible Knowing and Practice*, ed. Hilde Lindemann, Marian Verkerk, and Margaret Urban Walker, 1–20. Cambridge: Cambridge University Press.

Webster, George, and Francoise Baylis. 2000. Moral Residue. In *Margin of Error: The Ethics of Mistakes in the Practice of Medicine*, ed. Susan B. Rubin and Laurie Zoloth, 217–230. Hagerstown: University Publishing Group.

Widdershoven, Guy. 2005. Interpretation and Dialogue in Hermeneutic Ethics. In *Case Analysis in Clinical Ethics*, ed. Richard E. Ashcroft, 57–76. Cambridge: Cambridge University Press.

Widdershoven, Guy, Tineke Abma, and Bert Molewijk. 2009. Empirical Ethics as Dialogical Practice. *Bioethics* 23 (4): 236–248.

Wilson Ross, Judith, John W. Glaser, Dorothy Rasinski-Gregory, Joan McIver Gibson, and Corrine Bayley. 1993. *Health Care Ethics Committees: The Next Generation*. Chicago: Wiley Publishing.

Chapter 2
The Process of Ethics Consultation

This section describes the *Clinical Ethics Consultation* process, including the five stages: Pre-consult, Interviews, Mid-consult, Consult Meeting, and Post-consult.

Each stage refers to the *Ethics Consultation Toolkit,* which is included at the end of this book. The *Toolkit* provides detailed directions and tips for success, as well as sample forms and worksheets useful for various types of clinical consultation.

Ethics consultation, as described in this guide, is a five-stage process. Two of the stages involve all participants. The *Interview* stage introduces participants to the process, solicits their opinions on the situation, gets them thinking about what they see as at stake, and encourages them to consider other points of view. The *Consult Meeting* stage brings all participants together to deliberate and come up with a consensus decision on what is best to do.

For the ethics consultants, however, more is involved. Before the consultation process begins, consultants conduct *Pre-consult preparation*. After the interviews, consultants come together to do *Mid-consult analysis and planning*. And, finally, they follow up with a *Post-consult review*. Since this guidebook is written for anyone participating in ethics consultations, all five stages are described in order of occurrence.

> The *Toolkit* features a nine-page *Emerging Story Form* (see Appendix), which makes it easy for the consultant(s) to keep a detailed record of the ethics consult from start to finish, including:
>
> - The patient's clinical situation;
> - The patient's identity, including important relationships;
> - The patient's care team and their individual perspectives;
> - Family members' and loved ones' perspectives
> - The care teams, services, and/or programs involved in the patient's care, relevant to this issue;
> - System issues, including inter-team dynamics, policies and/or laws; and
> - The history of the consult request.

2.1 Consult Stage 1: Pre-consult

The ethics consultant will, from the start, be very careful with the information collected and with interview records. The personal and confidential information involved should not be casually shared or stored. Because some of the information may be gathered at first by persons other than ethics consult team members, appropriate measures are needed for confidentiality. These range from providing confidentiality training for administrative staff to implementing secure data handling procedures on computers. Access to files on the case must be subject to control. Furthermore, throughout the process, participants must be assured of this kind of confidentiality to encourage candid responses and discussions.

And yet, since the information gathering and interviewing may lead to a full consult meeting, participants should be aware, and should accept, that what they say to consult team members may become common knowledge to all participants. Where it is possible, one way of managing this sensitivity is to assure those involved in a conversation that information and opinions will be kept private, and that they will collectively decide what information can be shared and with whom.

The formal consultation process begins with the gathering of basic information. This gathering starts when the consult is requested, and can happen in a variety of ways. For example, someone might request a consult by contacting a member of the consult team. This team member could be one of the people who will actually do the consult or it could be an administrative staff person responsible for coordinating the consult process. Alternatively, those requesting the consult may be asked to begin the process by completing an online request. For example, they may have to complete a form available on the organization's web site.

> For a *Sample Intake Form*, see pages 10 & 11 in the Pre-consult section of the *Toolkit*.

The types of basic information that consult team members will want to record roughly correspond to the types of factual beliefs described earlier. The categories of information important to capture are:

- *Consult request:* Who has requested the consult? Which patient is the consult about? What is the reason for the request? Was there a particular event that raised ethical issues?
- *Patient contact details:* What is the patient's name, location, telephone number, other contact information?
- *Patient clinical details:* What is the patient's medical condition, diagnosis, prognosis, current treatment, cognitive status?
- *Patient identity details:* What are the patient's preferences, values, and beliefs? If the patient is not competent to participate in this decision, is there an advance care plan already in place?

2.1 Consult Stage 1: Pre-consult

- *Family members and loved ones:* Who is involved? What are the names and relationships of family members and loved ones potentially involved? What is their contact information?
- *Care team:* What are the names and positions of physicians, nurses, social workers, and other professionals either directly involved or knowledgeable about the case, and how can they be contacted?
- *System issues:* Are there any relevant policies or laws that apply to the case? Is there more than one service and care team involved, and if so, are there any cross-jurisdictional issues?

As well, the consultant would be wise to learn what people expect of the consultation. It may be more, or less, than what the service can offer. If this is not clarified upfront, the result may be frustration or disappointment on all sides. This type of outcome can be especially damaging at a time when ethics services are working hard to build their reputation for effectiveness. The questions to be answered are:

- What support do participants hope ethics consultation can provide?
- How promptly does the person requesting the consult wish to receive support?
- How extensive a consultation process is expected? A simple staff debriefing? An in-depth exploration of values with a family? An analysis and resolution of conflicts over priorities within the care team?

This is a lot of information, and it is useful to remember that proper understanding of these dimensions will only happen after many conversations. The *Intake Form* is a place to put these categories of information on the radar for people and to begin the early stages of exploring the story.

At this early stage, it will also be important to confirm that the consult request has been received and that a follow-up plan is in place. This is also a good time to help shape the expectations of the requester/requesting team, helping them understand how the ethics consultation service sees its role and what the process will look like. One way of achieving this is by sending a note, in the patient's chart, to the unit where the patient is to be placed. The note can acknowledge receipt of the consult request, set out the series of steps the service will intend to follow, and provide a point of contact should anyone have any questions.

> For an *Initial Response Template*, see page 12 in the Pre-consult section of the *Toolkit*.

With an early narrative summary of this information in hand, and initial communication in place, the consultants can begin planning the consultation process. **In the Pre-consult phase, three process planning issues should be considered.** First, how will consultant or consulting team members be selected? Second, how will information already collected be passed on to the team? Third, how will the next stage of information gathering be conducted?

Decisions on these issues are determined in part by the type of consult service available to the organization. If the resource is a single, professional clinical ethicist, the selection and transmittal process is much simplified. If it is a team or committee supported by a central office, the process is more complicated. Because of the variety of ethics resource models, the shapes of these plans will vary.

In the case of a consultation service using a team approach, it is very helpful to identify a consult lead. This person will be responsible for ensuring there are no cracks in the process–cracks that can easily be created in communication mechanisms when several people are working together.

In formulating plans, especially for a full clinical consultation, a good place to begin is a visit to the unit or ward and a review of the patient's chart. This is a valuable first step in helping a consultant become familiar with the setting where the issue has arisen. Sitting quietly and reading or writing at a table in a unit (and soaking in the atmosphere) can provide great insight into the culture of the unit. This can be invaluable in understanding the social dynamics at play in the relationships between the various people involved. Spending time reviewing a chart can also help a consultant to evaluate the information provided in the intake process. And of course, the chart itself is a rich source of new information, both about clinical aspects of the patient's situation and about the people involved in the patient's life. Finally, the chart review will indicate to consultants some of the individuals to be interviewed and suggest lines of discussion. In the context of a consult team, it is recommended that all team members take time to review the patient's chart.

At this point, consultants should develop a list of participants to be interviewed and a plan for gathering information (interviews, consult meeting, follow-up). This plan is then used in contacting and arranging interviews with prospective participants. If the consult is complex and involves many interviews, consult team members should probably plan to check in to share their learning and reshape the interview plan, as appropriate, based on feedback they receive.

> For a sample *Information Gathering Plan Form*, see page 13 in the Pre-consult section of the *Toolkit*.

After the consultants have an initial sense of the story and develop a plan for interviews and information gathering, it is important to communicate this to the person(s) requesting the consult and have this information placed in the patient's chart. This communication step shows that the request has been taken seriously and helps people understand the process being followed and the follow up to expect.

> **Resources in the Toolkit**
> The Pre-consult Stage of the *Toolkit* has directions, examples, and tips for success, as well as the forms mentioned:
>
> – **Sample Intake Form**
> – **Initial Response Template**
> – **Information Gathering Plan Form**

2.2 Consult Stage 2: Interviews

If the consult meeting alone could efficiently reach a decision, there would be no need for an interview stage. But experience has demonstrated that achieving effective outcomes depends on participants preparing themselves for *constructive deliberation*. The better the preparation, the better the decision-making. That is why the interview stage is as important as the consult meeting, if not more so. On occasion, a good interview stage can even make a consult meeting unnecessary. Take the case of Mr. Singh, for example.

> **Mr. Singh**
> Mr. Singh, age 70, was living in a long-term care facility. He recently experienced a severe brain injury and was transferred to the Intensive Care Unit of a local community hospital. Mr. Singh currently has minimal brain function, though he does breathe on his own. His ability to interact with his environment or his loved ones is severely limited. His eyes open from time to time and he seems to respond to his grandchildren's voices and to painful stimuli, but he's not able to relate beyond that. Mr. Singh is being fed by a nasogastric tube. Doctors have suggested that Mr. Singh's condition is unlikely to improve significantly.
>
> Mr. Singh has two children, Jane and Jeet. They are at odds about what Mr. Singh's treatment plan should be. Jane says she thinks any life is worth living and that her father should be maintained on the tube feed as long as his body permits. She also believes that medical science is advancing, and holds the very slim hope that he might actually get better. She loves her father deeply and does not want to lose him.
>
> Jeet, on the other hand, believes his father is suffering excessively. He says his father has no quality of life and would not want to be maintained in this state. Jeet, who seems to love his father as much as Jane does, wants his father to be free of this suffering and this kind of existence. He requests that the feeding tube be removed.
>
> Mr. Singh's care providers are similarly divided. Many members of his care team express that treatment should be discontinued for Mr. Singh–both because they think this type of life is no life at all, and because scarce ICU beds and resources should be used to help others who can be assisted more effectively. Some team members believe the treatment plan should focus on Mr. Singh's past known wishes; others believe the doctors should follow the wishes of Mr. Singh's family.
>
> To assist with the situation, the ICU staff request an ethics consult from the hospital's clinical ethics consultation service.

In the example of Mr. Singh, we set forth some of Jane and Jeet's understandings of the facts and values that may underlie their positions about what should be done for their father. We did not evaluate or judge their beliefs, but simply started to articulate and understand them. Similar descriptions will be made of the beliefs and values demonstrated by others involved, including members of the health care team.

At this stage, these descriptions are fallibly based on interpretation and speculation. We have not really listened yet, but have just tried to guess, based on our projections on the story. So we may have–most likely do have–mistaken assumptions at this stage. Really understanding people's perspectives requires conversation with them. Such conversations are not aimed at convincing people or trying to get them to see the world in a certain way. They involve reflecting back to people our understanding of what they are saying, to make sure we've got things right. Conversation also involves probing: asking questions about why people believe what they do, what the limits of these beliefs are, and what might make them change their minds. Getting an accurate description is not easy. Participants have to: (1) exercise moral imagination to "get inside" the perspectives of others; and also (2) rely on the evidence of behavior to date.

The articulation of perspectives and the effort to understand them is not a simple exchange of information, but rather a transformative process. Genuine listening and exchange of ideas changes opinions. In the process of putting ideas into words, we test out which words fit and which do not. And when we are encouraged to explain further what evidence we have for believing certain bits of information to be true, or how and why we value certain things, we are given a chance to reconsider and revise our views. As our ideas are taken up and bounced against counter-positions, they evolve. Sometimes, people don't change their minds but agree others' considerations are as important as their own. Either way, what emerges is a more fulsome articulation of what really should matter most in the situation–*effectively, the standards that a decision must meet if it is to be ethically justified.*

We'll get back to Mr. Singh in a moment. For now, I want to share an example of a time we were once asked to provide an ethics consult for a patient who was intensely ill in an ICU. He was being kept alive on a ventilator, and had to be sedated for the ventilator to be effective. If he was taken off sedation, he would have aspirated and died very quickly. The intensive care team shared the understanding that this was the only way to keep this patient alive. The team came forward and stated that this care plan was futile and they wished to withdraw the ventilator and let the patient die peacefully. The family insisted that the patient had strong religious views about the meaning of suffering in life and the idea of life as a gift from God. The family wanted the ventilator maintained to keep the patient alive, who they didn't perceive to be noticeably suffering.

During interviews in the ethics process, both sets of views developed such that people's positions changed. We asked the team whether maintaining the patient on the ventilator while sedating him would continue to keep him alive. They agreed that it would. We thus established that it wasn't really futile to continue to provide this care, if the goal was to maintain the patient's life in its current state as long as possible. The team agreed, and recognized that their disagreement was not about the technical effectiveness of ventilation, but rather about the goals of the patient's care. The team further reflected on what was more important to them–maintaining their commitment to a quality of life based on their own beliefs, or respecting the beliefs of patients in their care. Given the diversity of views among care team members and their commitment to respecting patient autonomy, the team felt comfortable having the goals of care guided by the patient's beliefs, as described by his family.

With the family, we explored the question of patient suffering. The family had said that the patient saw suffering as a meaningful part of life, to be avoided if possible, but endured if necessary. But then, in passing, the family also indicated that they hadn't perceived the patient to be suffering very much. This seemed out of place, so we inquired if they thought there was a point at which the patient would view suffering as not worth enduring, and would think death a better option. The family began to reconsider. After being reassured that our goal was sincerely to understand, and not to push them from their position, they admitted they weren't sure and needed to think more about this. They said they didn't know whether the patient had considered this possibility. They thought it was possible that had the patient not considered this previously, he would have wanted his family to decide. The family then said they might find death a preferable option to suffering at some point.

The dialogue we engaged in was shaping participants' thinking as they worked through the question of the goal of care. The descriptive work of listening was turning into the prescriptive work of developing shared beliefs and values to guide decision-making for this patient.

What the interviews are intended to do can also be seen by working backwards from a successful consult meeting at which a consensus decision has been reached and confirmed. Such a decision depends on participants collaborating productively to attain a shared understanding of what's at stake, what's most important, the options, which option is best, and how to apply the best option most effectively. This requires a very high degree of cooperation–especially among people whose disagreements and conflicts made the ethics consultation necessary. It takes more than one meeting to create this degree of collaboration. In fact, the reconciliation is set in motion by the interviews.

It is important to note that this reconciliation of initially opposing views is *voluntary*. The participants remain free to agree or disagree, to support or oppose. This requirement shapes the interviewing methods, and in fact the whole consultation process.

In general, the interviews cover all participants, who may fall into various categories: patient, family members and loved ones, care team members, system administrators or support staff, and, at times, external experts or consultants. In principle, everybody should be asked about every aspect of the situation in order to reveal differing perceptions of the situation and of one another. (The perceptions and values of consult team members, which may affect their performance as participants, are explored afterwards in the Initial Ethics Analysis.)

The interviewer in an ethics consultation serves as an active listener, with some ethics expertise, who seeks to deeply understand the interviewee's perspec-

tive and helps to contextualize the perspective against prevailing norms in society. The interviewer listens carefully, asks questions, probes opinions for the concerns behind them, and suggests alternate phrasings and expressions that are more neutral. The idea is to move beyond a particular position towards a deeper understanding of the participant's beliefs and values. The participant's underlying beliefs and values can then be more productively shared with others. The ethics consultant interviewer is not there to judge, persuade, criticize, or to tell their own story.

Interviewers remain explicitly impartial, merely trying to get clear on what participants mean. In so doing, interviewers share their understandings of values commonly held in society and ask participants how their views sit in relation to these ideas. In the process, interviewers also test possible translations of opinions that capture the essence of participants' perspectives in a form more likely to be accepted by others. In doing so, the interviewer influences the participant to: (1) reflect more deeply on what the real concerns are; and (2) become more aware of how to make those concerns more understandable to others who perhaps have very different concerns.

The interviewer should begin with some opening statements. There are some important things for the interviewer to keep in mind. First, because of the novelty of ethics consults, the interviewees probably won't be familiar with the consultant's role. Accordingly, any interview should begin by having the consultant clearly explain, in simple language:

- Who they are
- What their role is
- What the broader consultation process will include
- What information they are seeking to collect
- What will happen with this information
- What steps will occur after the interview

For patients and loved ones, it is important to clarify that the consultant is *not* a member of the care team, but rather a neutral party–who is, nevertheless, a specialist in helping individuals and groups work through the challenges of a disputed situation to reach an ethically justified decision. For care team members, it is important to emphasize that the consultant is an impartial external party who is there not to do the team's bidding but to help the team participate in a broader decision environment.

Consultants should also explain the ethics dimension of the situation at hand and illustrate how ethics processes can help with making hard decisions.

The interviewer should attend to each interviewee's understanding of facts, values and emotions. Consultants should attend to these three dimensions of the participant's experience and seek to understand his or her perspective on all three. In other words, consultants will want to distinguish and inquire into: (1) what the situation looks like to the participant; (2) what the participant believes is important to take into account; and (3) how the participant feels.

The challenge is that people may not be accustomed to making these kinds of distinctions. These distinctions may not be important for participants–what is most important for them is that their stories are deep and accurate, and shared with relevant others. Interviewers will have to be good at mentally making these distinctions while listening to interviewees' stories.

A broad opening question can often provide an entryway for interviewees to begin sharing their perspectives; the interviewer can then interrupt from time to time to review and reframe what she has heard and probe further to allow a deeper understanding to emerge. Possible opening questions include: "What is your perspective on what is happening here?" and "How do you see this issue?"

Beliefs about the world form the descriptive backdrop of the story. Talking to interviewees about their understanding of the facts is about trying to understand their view of the world. For example, in Mr. Singh's case, it is important to ensure both Jane and Jeet have a good understanding of their father's medical condition, including his diagnosis, possible treatment options, and implications.

Examples of good questions for understanding facts include:

- What is your understanding of the situation?
- How did we arrive at this situation?
- What else do you need to know?
- What else would you like to know?
- If this situation continues unchanged how do you think it will affect...

  You?
  Your family/team?
  Others involved?
  Society at large?

Notice that none of these questions involve judgment–they are all invitations to share a perspective. This is important because the goal of the interview is not to judge; it is to try to understand. But trying to understand does not mean that the interviewer should not question a perspective. Often people (including me and you) have many beliefs about reality that we don't even recognize. These beliefs arise because of stories we have made up about people and events without even realizing it. So as interviewees begin opening up and sharing their understandings, it is entirely appropriate to try to explore with them where their ideas have come from.

Examples of helpful probing questions include:

- How did you come to this understanding?
- What examples are you thinking of that led you to this concern?
- What makes you think so?
- What evidence are you relying on?

Acknowledging emotions enables collaboration. It is an understatement to say that emotions are a very important part of one's experience. We bring our hearts to what we do. In a health care related situation, our emotional reactions can be stronger than usual because of the gravity of what is at stake. Our emotions are heightened even more when the situation becomes one where uncertainty or disagreement has emerged. Therefore, central to people feeling heard and taken seriously, is having their emotions acknowledged and, where appropriate, supported. So the interviewer will have to pay close attention to what feelings (as opposed to thoughts) the interviewee is experiencing in their heads, their hearts, and their bodies. This is not only an intellectual exercise; it is also an emotional one. In other words, the consultant will have to try to see the world through the interviewee's eyes, and then allow their hearts to be open to get a sense of what emotional impact this view of the world might be having on the interviewee. This may sound like a tall order. But if the purpose is to understand, then emotional investment is required.

It is one thing to try to understand what someone might be feeling and quite another to try to frame those feelings in a language that interviewees can relate to without feeling threatened. Part of the interviewer's challenge includes sharing how she thinks the interviewee has been impacted emotionally through the experience of the situation under discussion. If not handled well, this can leave the interviewee feeling judged and, worse, feeling perceived as irrational. Language such as "I imagine you must be feeling frustrated by all of this" or "This must make you feel sad," can be useful to help the interviewee know what the interviewer thinks the experience must feel like. It can also help legitimize the interviewee's feelings, seeing them as a reasonable consequence of experiencing the world as the interviewee has.

There are also times when the person being interviewed is so upset that it can be helpful to acknowledge their emotional state upfront and try to assist them to come to a place where they can proceed with the interview calmly. Here, language such as the following might be helpful: "I get that you are really upset," or "I know you feel very strongly about this," or "This must be very difficult for you," followed by "I really would like to understand your perspective better … could we take a moment and then could I ask you to share what's going on for you?"

Additional language for acknowledging, exploring and debriefing emotions might be:

- What is in your heart as you go through this?
- How are you feeling about this?
- You seem very...
- Are you feeling...?
- I'm sorry you have to go through this.

Values are *what matters* in the situation. A crucial purpose of the interview is to allow for a deep understanding of what the interviewee sees as important in the situation. The better this objective is met, the more likely that a solution responding to these considerations is likely to surface. Indeed, if we don't understand what is really at stake for participants in the consult, we most likely won't be able to meet their needs.

When talking about values, instead of using vague terms like 'fairness' or 'integrity,' it is helpful to be specific about what matters in a situation. These specific statements can be thematized at a later stage under broader value headings.

Our values can be important to us intrinsically or instrumentally. As described in the first section of this guidebook, something is *intrinsically* or inherently important to me if it is important for its own sake. Something is *instrumentally* or strategically important if it gives me other things of greater value to me. The ability to distinguish between intrinsic and inherent values and discern what is at stake at a deeper level in our everyday choices is important for several reasons. First, instrumental values don't provide direction for how we should behave or the choices we should make. Second, having clarity about our intrinsic values will not only allow us to find our way through difficult decisions, but will also allow us to respond directly to what is really at stake in the decisions we make every day.

The interviewers should help interviewees make explicit what matters to them in the situation. When a consideration is identified as important, the interviewer should explore whether it is important for its own sake or because it provides or leads to something else of greater importance. If the latter, the consultant should capture both the instrumental and intrinsic value.

For example, if it is important to the nurse manager that "all team members clearly chart conversations with a patient about what is important to them while they are in hospital," the consultant should explore *why* this matters. Is it because it is important to the nurse manager to "minimize exposure to legal liability," or "ensure consistency of care," or "best respect the autonomy of the patient," or "assist family to understand the perspective of their loved one"? In having conversations about values, the consultant should avoid one-word values because these are open to broad interpretation. The purpose of the interview is not to pin the interviewee to a value, but rather to develop a fine-grained sense of what really is at stake for them.

Examples of good open-ended questions for understanding values include:

- What is important to you as we move forward?
- Why do you prefer this solution–what does it give you that you believe is important? What would a solution have to achieve for you to be happy with it?
- In our society, this value (e.g., equity) is important–what is your sense of this?
- What do you think this value means in our context? How important is it?
- If someone who would disagree with your perspective were here, what would they say is important?

Good probing questions for helping to deepen the interviewer's understanding of what matters include:

- Why is this important?
- Here is a competing value (tell a story); how would you balance these two?
- What would have to happen for you to change your mind?
- What does this tell you about what else matters to you?

Framing the role of the interviewer. The consultant should be transparent about the purpose of the interview. For example, early in the interview, explain: "My purpose here is to really understand your perspective as best I can. So I would like to ask you to tell your story. I may interrupt from time to time to make sure that I've correctly understood what you're saying, and I may also ask questions about your view of something. But please know that I'm not judging as I do this; I just want to get a deeper sense of what you see going on and what is important to *you*."

The consultant must be sensitive to interviewee's emotional reactions to the interview. Sometimes the interviewee will react as though questions or suggestions reflect the consultant's views. It will be counterproductive if the consultant's values and understanding are perceived to challenge or be in tension with those of the interviewee. Such an interview will have an antagonizing, rather than supportive, effect. The interviewer must take pains to show they have no intention to change minds or discount opinions. Finding the solution to issues is not up to the interviewer; it is up to participants in the consult meeting, who must talk together and share their views. The interviewer helps get that conversation underway and keeps it moving along. For participants, the value of the interview is to receive the reactions and suggestions of an impartial third party in order to improve their own understandings of issues and their effectiveness in communicating their beliefs and feelings to others.

It is actually fairly easy to tell when people are asking questions because they want to better *understand* a perspective versus when they are asking questions because they *disagree* with someone's view. Again, the purpose of the interview is to *understand*, so the ethics consultant will have to be good at self-regulating and remaining vigilant as far as not becoming judgmental.

While examining perceptions, emotions, and values, the interviewer will let the interview take its own course, leading or proposing rather than directing or controlling. In this way, the interview will evolve to feature topics the participant feels most strongly about. At the same time, participants will learn the importance of neutrally presenting their views, discover possible misunderstandings, gain better understanding and respect for the consultation process, and become better prepared to participate in discussions. Such an approach is consistent with fostering participant trust and openness for the consult meeting to come.

Finally, because the interviews are likely participants' first exposure to an ethics consultation, the way the interviews are conducted will significantly impact the reputation of the consultant and the consult service. To ensure the credibility of the individual consultant and the service, the interviews must be empathetic, thoughtful, and systematic.

Interviewing care team members. Care team members often see things from two perspectives. First they have expert opinions about patients' clinical situations, and about families and family dynamics. Second, they have personal opinions on available choices and on what the right decisions might be. It is helpful to learn about both perspectives and how strongly they are held.

An important area for care team input involves patient competence. Is the patient able to make thoughtful choices and decisions? If the patient is not able to fully understand the situation or the consequences of the decisions, to what extent is the patient able to participate in the decision-making process? If the patient is deemed unable to participate in decision-making about care choices, on what evidence is this determination being made? Has the patient made some kind of advance care plan or personal directive?

There may be disagreement about some of these questions among team members. As well, assertions about the patient may be made without clear and convincing evidence. The consultant must help all those involved identify any assumptions being made for which the justification or reasoning is either unclear or lacking.

> **Resources in the Toolkit**
> *Elements of Good Facilitation*, found on page 4 of the Toolkit, suggests questions for exploring facts, values and emotions.

2.3 Consult Stage 3: Mid-consult

When the participant interviews have been completed, but before the consult meeting occurs, ethics consult team members should reconvene to undertake two important tasks. The first is the initial ethics analysis, where team members share what they have learned and process all this information. The second is planning next steps in the consult process.

2.3.1 Initial Ethics Analysis

In undertaking an initial ethics analysis, team members bring together all the interview information they have gathered and sharing their views and insights. They perform their own preliminary analysis of the main ethical issues to be dealt with in the consult meeting. In doing so, ethics team members should identify and come to terms with their own beliefs, values, and emotions. Basically, they interview themselves and one another and bring themselves to a procedural parity with all the other participants.

The process for this initial ethics analysis includes the following steps:

1. Clarifying the key question
2. Assessing the facts
3. Considering guiding values
4. Identifying possible communication gaps
5. Identifying the need for emotional support
6. Exploring possible solutions

> **Resources in the Toolkit**
> The Initial Ethics Analysis of the *Toolkit* provides methods and worksheets for undertaking these steps.

The ethics consult team would begin with Step 1, clarifying the key question. In most situations where ethics support is requested, the contexts are very complicated. There are often many concerns embedded in the situation: patient-family relationship issues, intra-team relationship issues, inter-team relationship issues, and broader system issues. Step 1 involves clarifying exactly what is the problem, area of tension, or unanswered question in this particular situation that the consult should try to address.

Here the consult team recognizes that the type of question asked in this consult exercise will determine the type and scope of the answer that will be arrived at. The team will thus want to ensure that group members are all working on the same problem and asking the right question to help solve that problem.

Sometimes more than one legitimate key question will emerge in a consult. The challenge in these cases is to decide which question(s) should be prioritized for addressing in the time available in the consult meeting. In effect, this is about setting the agenda for the consult meeting. Consistent with the entire Ethics Analysis, the point of this step is not to decide this question unilaterally. Rather, the consult team should reflect on what the key questions are for the various participants in the situation, whether there is a central question that is at the core of the situation and of interest to all, and how this question should best be phrased. The consult team will rely on this analysis in the early part of the Consult Meeting as the agenda for that meeting is established. To help illustrate the initial ethics analysis process, let's consider the situation of Arthur:

Arthur
Arthur attends an adult day program. He had a stroke that has compromised his ability to swallow. Arthur's judgment and decision-making processes seem to be affected and he shows some signs of brain damage. Arthur is very overweight with a distended, protruding abdomen.

One day staff and a dietician observed him coughing frequently during his regular-textured meals. The team believed it would be safer for the client to eat soft-textured meals. For this reason, Arthur was offered pureed and minced food at the next mealtime. When Arthur received the meal, he became very upset. He used harsh language to describe the meal and he became angry with the staff.

Arthur's wife, Missy, was notified about the swallowing issues. The day staff expressed their concern that if Arthur continued to eat regular-textured meals, it would put him in danger of choking. Because the Heimlich manoeuvre would be challenging to do for someone his size and weight, this could lead to Arthur's death.

Missy indicated that she too had observed Arthur's difficulty swallowing. Indeed, they had an emergency situation happen at a restaurant. Nevertheless, Missy believes they should cater to Arthur's desire for regular-textured meals as he has always loved food. Missy has said to let him die choking if things came to that.

The Day Program requested an ethics consult and the consult team has begun to provide support. They've consulted the chart and met with several key people. They have now come together to take on the Mid-consult steps found in the *Clinical Ethics Consultation Toolkit*.

2.3 Consult Stage 3: Mid-consult

In Arthur's case the following questions will likely arise for consult participants:

- Should we let Arthur continue to eat regular-textured foods?
- What does Arthur want?
- What is the professional obligation of caregivers to Arthur?
- What should the goals of care for Arthur be?
- Should health care providers ever help residents do what they consider harmful?
- How should the team resolve its internal disagreement?
- How should the team respond to Arthur's desire to eat regular-textured food?

In assessing these questions, the consultants should recognize that some questions are really about missing information (e.g., What does Arthur want?), while others are about how to deal with an issue (e.g., What should the goals of care for Arthur be?). It is the latter that they will want to choose at this stage.

In framing the key question, the team will want to avoid questions that yield "yes" or "no" responses (e.g., "Should health care providers ever help residents do what they consider harmful?") in order to allow a broad range of answers to emerge. Questions that begin with "What" or "How" work well. It is also important to pose the question in neutral terms and limit descriptors to those about which there is explicitly shared agreement. Nor should the question predetermine the answer. It can be useful to focus on a broad question that, if answered well, will likely include more specific questions and provide meaningful direction for moving forward.

In the clinical setting, a perennial problem I run into is that a shared understanding of the goals of care for a patient is not in place. Too often, the goals of care have been prescribed by the patient's physician or physicians without sufficient consultation with professional colleagues or with the patient herself. So usually a very good question to start with is: "What should the goals of care be for this particular patient?"

In Arthur's story, however, there does seem to be a clearly defined problem for which the team seeks a solution. The question is: "How should the team respond to Arthur's desire to eat regular food?" I'll use this as the key question for the remainder of this illustration.

The consult team would then move to Step 2, assessing the facts. As discussed earlier, how one answers the key question depends in part upon one's perspective. The ethical justification of a solution will be determined by the extent to which the consult analysis is based on a shared and accurate understanding of reality.

Step 2 is about assembling various participants' pictures of reality into a shared understanding of the landscape to ensure that decisions are based on the best evidence available. It is about painting a broad picture, based on the interviews and chart reviews, and then identifying and addressing overlap, divergence and gaps.

There are a number of categories of factual information that a good ethics analysis will require. In Arthur's case, the categories of information we will want to include are:

- Arthur's clinical condition
- Arthur's identity–who he is as a person
- Who knows Arthur well and can discuss his values and beliefs
- The impact of Arthur's choking on him
- The impact of Arthur's choking on his family
- The impact of Arthur's choking on his caregivers
- The impact of Arthur's choking on those around him in the Day Program
- In Arthur's life, the relative importance of food, and of minimizing suffering and risk of death
- Technical details around Arthur's eating
- Possible ways of preventing Arthur from choking
- Support and training available to staff for responding to choking clients
- Staff history in dealing with and supporting such clients
- Organizational policy for responding to client emergencies, including choking

At this stage of the analysis, the consult team should ensure that what gets listed here are beliefs about the world (things that are true or false), and *not* values (what is important to us). Beliefs about the world usually involve declarative sentences with some form of the verb "to be" (e.g., "Arthur has an eating disorder." "Arthur is at risk of choking if he eats solid foods."). The goal is to ensure that the team making the decision is on the same page, looking at the same picture. List only those beliefs about which it is important that everyone agree, including things that may be contentious but relevant.

It can be helpful to assemble reported beliefs about reality into three types:

1. Things we know for sure. These are agreed upon facts, which we have good evidence to justify. In Arthur's case this may include information about Arthur's physical situation, diagnosis and prognosis.
2. Things we are unsure about, but can figure out. This is information that is contested or currently unjustified, but for which evidence is likely available. For example, there may be agreement that Arthur wants to eat solid food. But we may not know the meaning of food for Arthur, what about eating he finds meaningful, and other aspects of Arthur's values around food. Hopefully this information will have been uncovered with careful interviews, but sometimes these gaps are not identified until the Initial Ethics Analysis stage. When such gaps are identified, it's also useful to plan who will do the research required to fill them in.
3. Things we don't and likely cannot know. This is information that is missing and for which there is no evidence available. For example, it may be useful to know what happened in the emergency at the restaurant, which Missy referred to. But Missy and Arthur may not want to talk about it. Or it may be useful to know Arthur's perspective on the possible pain and complications he might experience from having staff work on him if he is choking, but he may not have the capacity or willingness to offer this perspective.

2.3 Consult Stage 3: Mid-consult

After assembling the facts it is time to consider guiding values.

In step 3, the team lists and reviews what is at stake for participants in the situation. The team should consider different participants' priorities, areas of overlap and agreement, and areas of tension.

In Arthur's story, the interviews may have revealed the following list of important considerations:

- We discharge our professional obligation.
- We treat our clients with respect.
- We make decisions in a manner consistent with Arthur's values and beliefs.
- We maximize the time we spend in direct client care.
- We minimize exposure to legal liability.
- We minimize exposure to professional censure.
- We minimize health care costs.
- We respect the values and beliefs of families.
- We treat our colleagues with respect.
- We respect the professional autonomy of our staff.
- We minimize harm to our patients, residents, and clients.
- We act on good evidence.
- We demonstrate compassion to our residents.
- We support the relationships of people living in the facility.
- We meet the needs of all of our residents equitably.
- We support those residents who are particularly vulnerable.

In addition to the values and beliefs of all other consult participants, there is the ethics consultants' own beliefs about what is true and what is important. The consult team members must reflect on their own beliefs and values as part of this descriptive work. Such reflection is important, because one's own value commitments could unconsciously influence one's response to various people involved in the situation. If consultants are aware of their own values and beliefs, they are then able to promote integrity–that is, living intentionally and deliberately.

Thus the process should make room for this self-reflection to be explicitly undertaken. There are several ways of probing this. Team members can ask each other, "What would you do in the situation?" or "What do you think the right answer is here?" Team members can also ask each other and themselves, "Whom do you find yourself relating to in the situation?" And for all of these questions, they can further ask, "Why?" The purpose of this self-reflection is not for the consult team members to decide what should matter most or prejudge the solution. Rather, it is for team members to be mindful of their own biases. To this end, it can also be helpful for team members to name out loud what their own predispositions are, where these are coming from, and why they need to be careful to keep these in check as the consult proceeds.

Having identified which priorities are shared and which are in tension, the consult team members can plan for a number of things: Which participants need to hear what perspectives in order to be able to consider the situation in full? Whose thinking is incomplete and which individuals should be supported to think more carefully about how their perspectives are justified within the broader norms of society?

In Step 4, after a careful review of participants' perceived facts and values, the team identifies key gaps. These may become more apparent at this stage. For example, there may be technical information that patients or family members are missing or have wrong. Or members of the care team may not really understand what is at stake for family members and why. It can be very useful for team members to identify information gaps at this stage and to develop strategies to fill in such gaps *before* bringing everyone together in a larger meeting. At this step, consult team members need to develop a plan for what conversations should happen with whom before a broader meeting is convened.

The team then moves to step 5: considering emotional support. When reflecting on the situation from a distance, consult team members may recognize that certain participants are really struggling emotionally with the issues, such that they aren't able to participate in the shared reasoning process required by the consult.

For example, it may emerge that Missy is having a very hard time with the whole situation and that she is feeling distraught, frustrated and angry. Her feelings may be coming from a place where she does not share Arthur's perspective in all of this, but is finding it too difficult and painful to think about and is just saying what she thinks will placate him. The ethics team may decide to explore whether social work could provide some support to Missy to help her get through this. Or they may suggest to Missy that she might wish to consider such assistance.

Step 5 is an opportunity for consultants to think of resources that might help individuals.

In step 6, consult team members explore possible options for answering the key question. This step is aimed at creatively exploring what kinds of things, conventional or not, might meet the criteria indicated and answer the key question. In an exercise similar to one conducted at the larger meeting, options should just be listed and not judged. The team should remember that just because an option is named does not mean it will be put into action.

In Arthur's situation, the possible options that may have arisen in the interviews include:

- Only provide Arthur the puréed foods recommended by the dietician.
- Support Arthur to eat what he likes and provide whatever assistance he requires.
- Have Arthur sign a waiver as a way of helping staff confirm his wish to indulge in this risky behavior.
- Enable Missy to feed him the food he likes, but do not participate in feeding Arthur solid foods.
- Have Arthur start with a puréed meal to meet his nutritional requirements and then allow him to have regular-textured food that he particularly enjoys to end the meal.
- Support Arthur to have one regular-textured meal a day and assist him as required, and then provide puréed foods for the remainder of his diet.

- Support Arthur to eat what he likes and provide whatever assistance he requires, but only when there is appropriate staff available to assist should he aspirate.

Again the purpose is not to prejudge a solution, but rather to begin to anticipate possible solutions that may emerge, or that the team could introduce at the appropriate time to help stimulate other ideas from the group in the consult meeting.

2.3.2 Planning Next Steps

After the initial ethics analysis is complete, members of the team should determine the next steps they will recommend in moving forward with the ethics consult. These will include:

- any actions required regarding communication gaps
- any actions required for enabling participants to get needed emotional support
- planning the logistical and organizational dimensions of a larger consultation meeting
- thinking through strategies to make the consult meeting move forward efficiently
- drafting the consult update report that will be shared with participants and put on the chart at this stage of the process.

> **Resources in the Toolkit**
> The *Toolkit* provides a *Consult Meeting Planning Worksheet* on page 23 to assist with this planning.

Consult team members should begin by thinking about who should be involved in a consult meeting, what the objective of the meeting should be, where and when it should take place, and any other relevant details. Team members should then carefully plan and undertake the steps needed to achieve this.

If possible, the following people should be present at the meeting:

- those with the authority to make the decision
- anyone who will be impacted by the decision
- anyone spoken about in the consult meeting

It's essential to ensure that those present have the authority to settle the question at hand. This affects the development of the agenda because there is little point in discussing a question or topic unless meeting participants between them actually have authority to make the decision being discussed. For example, if the question involves the goals of care for a patient who is competent to participate in decision-making, an ethically justified decision cannot be made in the patient's absence. In short, if certain key people do not attend, the meeting will have to be called off or redirected into a kind of preliminary meeting lacking final decision-making powers. Probably not everyone affected will want or be able to participate in a consult meeting. However, for people who will be impacted by the decisions at hand, it is important to ensure that there is an appropriate place for their perspectives to be heard somewhere in the process. (For example, they could be interviewed and their perspectives could be shared via a consultant.)

The consult team should choose an appropriate time, location and type of room for the consult meeting. The room itself should be big enough to comfortably accommodate all participants, as well as flip-charts or any such meeting aids the consult team might need. Team members should think carefully about patient needs and ensure that wheelchairs or even beds can be accommodated if appropriate. If a patient is unable to leave his or her room, it is entirely reasonable to consider meeting in the patient's room or unit. The patient would need to be comfortable with the space being used in this way.

Ideally, the meeting will take place at a location, date and time that is mutually acceptable. This can help to acknowledge and mitigate power imbalances. In reality, members of the care team and ethics service may have limited availability. In this case, the consult team should identify strategies that will assist patients and families to feel as comfortable as possible. This will vary depending on setting, but simple things, like calling everyone by their first name, can assist with levelling the power in the conversation.

Once everyone understands the purpose of the meeting and how it will proceed, the team should develop a detailed list of items needed to begin the plan. The list will clarify who will invite whom to the meeting, who will book the room, and so forth.

> **Resources in the Toolkit**
> For the Mid-consult Stage, the *Toolkit* contains resources to help with the *Initial Ethics Analysis*, including clarifying the key question and making explicit the facts and values. It also includes a detailed *Consult Meeting Planning Worksheet* on page 24.

2.4 Consult Stage 4: The Consult Meeting

This stage includes preparing for and conducting the consult meeting.

The purpose of the consult meeting is to bring together participants and enable them to listen to one another and carefully reflect on what's most important in the situation, as well as to support them to collaborate and build a solution that allows all impacted to live with integrity. The consult meeting is key for making all of this happen. Depending on when in the process the meeting is taking place, it could be focused on any or all of the following: identifying issues and the key question; creating mutual understanding and developing a larger, shared view of the descriptive background; creating understanding about what matters to each party and moving towards a shared understanding of what should matter most; brainstorming solutions; and/or developing a solution that best lives up to what matters most.

It is conceivable that all of this could happen in one meeting if the issues are not too complex, if the emotions are manageable, and if the disagreement is not too great. For more complex situations, it is likely that only some of the above goals are reachable in a single meeting, and multiple meetings will likely be required to complete the process.

The consult meeting brings to fruition the dialogues begun in the interviews by bringing everyone together to make the essential decisions. This meeting is crucial to the achievement of an ethically justified decision. The ethics consult team approaches the consult meeting with three goals in mind: full participation, developing the best possible solution, and gaining full understanding and acceptance of the decision. To achieve this tall order, the ethics consult team brings a decision-making procedure and a set of tools and skills to the consult meeting to make it work.

At this planning stage, the consult team will want to review the ethics decision process that will be brought to bear at the meeting and ensure all team members are on the same page about the specific steps to be followed in the process. Additionally, the team will want to reflect on such practical matters as who should sit where, what the respective roles of consult team members should be, what introductory remarks will be made by whom, and what opening statements will be solicited.

Seating arrangements should be considered. This can be a complicated matter and there is no one right way of doing things. However, there are a few considerations that might help. First of all, it shouldn't be left to chance; if seating arrangements are not attended to, facilitation could be more challenging, making a positive outcome more difficult to achieve. Particularly vulnerable participants would benefit from have someone supportive sitting by their side. That person could be a family member or friend, or a member of the consult team. Separate participants who may be hostile to one another might benefit by having a member of the consult team sit between them. However, it may not be a good idea to put opposing parties at opposite ends of the room, as this may highlight the tension between them. Members of the ethics team should disperse themselves amongst the group in order to spread out control of the conversation, and avoid the sense that the consultants are all one team with one voice. This also facilitates eye contact and other forms of communication amongst the participants.

Consult team members will play several roles at the meeting. These include: meeting chair, process facilitator, active listener, scribe, timekeeper, and sometimes intervener, when emotions run high. These are challenging roles and it helps to divide them amongst the team members. How roles are divided depends on the number of team members, their respective experience and strengths, the number of people in the consult, and the type and extent of tension in the situation.

Parts of the meeting will probably involve highly emotional expressions from some participants. Our goal is not to eliminate emotion; our goal is to facilitate participants' understanding of and respect for how someone is feeling, and what is at stake for that person. The ethics consultant asks participants not to interrupt each other, to try to open their hearts and minds, and to demonstrate mutual respect. Mutual respect is necessary not only to honour personal dignity, but also because it is the most efficient and possibly only way to achieve a successful outcome.

The last part of planning the consult meeting is to document the process. Documentation should include the steps that have been followed to this point, summarizing the key question, salient facts, basic values at stake, and the next steps in the ethics process.

> **Resources in the Toolkit**
> Specific directions and suggestions for preparing and conducting the Consult Meeting can be found in *The Toolkit* on page 25.

2.4.1 Conducting the Consult Meeting

This section describes the consult meeting and the tools and skills required of the ethics consultant(s). The ethics consult team will need to establish the proper atmosphere and decision-making procedure. They'll need to identify issues and build a shared meeting agenda. They will need to explore beliefs about facts and values, brainstorm options, and guide the building of a solution. Finally, they'll need to review the solution to ensure it lives up to key values and then set out next steps for moving the solution forward. If the consult ends before a decision is reached, they will need to establish where in the decision-process the group is at and how the group will continue to engage in the process, assuming it wishes to do so.

The ethics consultant or consult team has a special role in maintaining a proper atmosphere during the meeting. To do this, it helps to keep several things in mind.

First, a systematic ethics analysis of an issue in a collaborative setting will require everyone involved to think through the relevant facts and values in the situation. Taking the time to understand how people are feeling can help the consult move forward. If participants are feeling upset and have not had their emotional state or its cause acknowledged, this may prevent them from being able to undertake the reasoning exercise that is required. Our emotional states contribute significantly to making ethically justified decisions and should be acknowledged and addressed in the ethics process. Part of demonstrating respect for another involves opening one's heart and trying to empathize with what the other might be feeling.

Second, although the consultant has a certain leadership role in explaining, guiding and initiating, the meeting really belongs to the participants as a whole. Once a section of the discussion is under way, the consultant stands slightly to the side as a facilitator. The facilitator watches for non-verbal cues and notices how participants are feeling. He or she acknowledges people's emotions, tries to maintain balance in the conversation, and keeps an eye on the clock.

Third, the problems and difficulties being grappled with are not those of the consult team. The ethics consultants own the process, but they don't own the problem, or the solution. Our job is to provide opportunities and tools in a transparent way. If we do our job right, we open the way for participants to do their job: the hard thinking and decision-making.

The decision-making procedure employed by the ethics consultant follows a general pattern: (1) naming issues, (2) identifying values, (3) brainstorming options, (4) building a solution, and (5) confirming the decision and way forward. These steps take us from what participants see happening (issues) and what they think is important in making a decision (values), through the alternatives available (options), to a preferred and enhanced option (solution), and finally to a critical review of all the steps taken in order to agree that this is the best decision (confirmation).

The ethics consult leader should provide an opening statement that explains these steps (even if they have been explained before), emphasizing that the use of a formal procedure like this ensures that everyone gets to have their say, and see how their views contributed to the eventual decision. A handout outlining the steps involved will help participants follow the progress of their deliberations.

> **Resources in the Toolkit**
> The *Toolkit* provides a *Consult Meeting Data Capture Form* where the consultant can capture what emerges in the consult meeting.

A good place to start is to have participants state *the key issue or issues* in the current situation. The issue is a broad phrase, statement or question that describes the problem area. Difficult situations tend to present many issues that deserve attention: relationship issues amongst and between family members and care team members; broader resource allocation issues; questions of appropriate standards of practice; legal tensions; and the list goes on. Such complexity is typical of cases in health care today.

For this reason, it is important to clarify which are the burning questions for everyone involved. A key question is a statement of a specific problem raised for discussion and resolution. A useful guide to finding the *key question* in an issue is to ask what specific problem statement needs to be answered at the meeting in order to move the situation forward significantly.

Since we are beginning with a list of issues, participants will quickly see that they should try to turn their wishes into neutral statements. For example, an emotional family member may say that he "doesn't want these guys to cause my father any more pain." This could become the issue of "controlling and reducing pain," thus giving the family member a neutral vocabulary with which to speak to the care team. Or, a different ethics consultant might make a different suggestion: "So the issue is: what's the best care plan for your dad, and how can that care plan manage his pain appropriately?" This wording introduces a distinction between the issue, which is a care plan for Dad, and the value, which is to minimize pain.

During this discussion, areas of misunderstanding or disagreement on the facts may be identified and sometimes resolved.

The results of this discussion will either be a list of issues to be tackled or an emerging understanding within the group that there is one central or core issue to be resolved. If there is a list of issues, the next step is to go through the list with the group and prioritize: "Which do we really need to tackle today, or which do we need to tackle before we can get to some of the others?" For issues that are deferred, the consultant should try to identify how and when these might be addressed.

Once there is a clear issue to resolve, the consult meeting moves on to identifying the values in the perspectives of all participants. To do so, the ethics consultant can say something like: "Now I'd like to hear what's important to each of you in connection with what the goals of care for Dad should be." Make it clear that at this point we're not launching a discussion but just seeking *each person's* point of view.

The crucial thing here is to go behind previously expressed demands to find the personal values animating requests. In mediation language, this is referred to as 'getting from positions to interests.' The idea is that, on a given issue, a person may have taken a position: "I want Dad to get good pain medication. I want Dad to get this pain medication." Such positions are attempts to express in practical terms the person's deeper values by advocating a solution that would appear to satisfy those personal values. Difficulties arise when a proposed individual position conflicts with the positions of others. What the consult meeting is trying to do is to *find a restated consensus position, built on the foundation of the values underlying the individual positions.*

There may be many practical ways to realize particular values, and some alternatives may be more acceptable to other participants. By going back to the underlying values, we open the possibility of alternate measures not initially thought of, but perhaps better able to meet the deeper wishes of other participants as well. *Mediation, therefore, uses methods of questioning, probing and reframing to create neutral, acceptable or more positive formulations of values, in order to find common ground behind apparently conflicting positions*

Value-probing conversations encourage reflection and deepen people's understanding of the meaning in their lives. Along with this may come a better understanding of how much we value other people and their connections to us. It isn't only spouses or family members who gain a new appreciation of the value of others; the same is true for health care providers as they learn more about patients and their families. The ethics consult meeting is a learning process in which each participant gains a better understanding of others' beliefs and values, and perhaps finds their own beliefs and values modified as a result. Ideally, all participants find themselves moving away from pre-formed opinions towards a more shared understanding of the question to be resolved.

In pursuing these discussions, it helps for the ethics consultant to use a flip chart to list values proposed by participants. Through several revisions and re-phrasings suggested by participants thinking together, a list is compiled of shared, neutrally stated values capable of being used as decision criteria to assess alternate solutions. In some situations, it helps if the participants can work out an agreed-upon ranking of the values, but priorities among them are not necessary.

2.4 Consult Stage 4: The Consult Meeting

To keep everyone oriented, the ethics consultant maintains continual awareness of the meeting's progress through the various steps and of the significance of what is being discussed at every moment.

The meeting then turns from what people want and value to brainstorming options, exploring what is possible and available. Naturally, the health care team members participating will play a central role in compiling a list of alternate methods of treatment that represent different medical approaches or even philosophies. But often an option will be initially suggested by a non-medical participant. As all participants will by now understand, this list will represent the broad strategic choices from which they have to decide.

The ethics consultant will explain that once a preferred alternative has been chosen, it will later in the meeting be adapted and enhanced in detail to make it as effective as possible. But this beginning stage has to emphasize the broadest range of general directions, so that a strategic choice becomes possible. It is especially important that the range of possibilities be comprehensive and unrestricted, in order to prevent a later objection that a particular strategy was not even considered at all.

When a list of strategic options has been set forth, the group then needs to evaluate them to build a solution. The idea is to use the list of values to rank the options and reveal the most acceptable one. In practical terms, I suggest using two parallel flip charts, or two pieces of paper of which everyone has copies. One page lists the values; the other lists the options. Then we take each option and compare it against the values one by one, asking, "How does this option fare against this value?" Possibly we can use a simple numeric or "report card" method of ranking the options against the values. By the end of this exercise, participants are beginning to have a good sense of how each option fares against what values are most important to participants in the situation.

Usually, one option begins to seem better than the others. Sometimes two options seem tied. When that happens, we can call upon relevant expertise to determine whether we could construct a single preferred direction combining the merits of both strategies. With a preferred option selected by the group, the most difficult steps in the consult meeting are finished. By this time, participants have learned how to work together, and things start moving more quickly.

During the evaluation of the options, some of the rejected options probably revealed certain merits not possessed by the selected one. We don't want to lose those benefits if we can capture them somehow. Moreover, the chosen option has to be designed in practical detail to take full advantage of the medical resources on site or in the community.

Once a solution has been tentatively worked out in adapted detail, the final step is to confirm the decision and the way forward. This begins with a critical review of the previous steps that have led up to the decision. In some cases, where the process has succeeded in getting all participants on side, the participants may decide that this review is unnecessary. But until this point is reached, and especially where there is some opposition or uncertainty, it is very important to retain the review step.

There are several basic reasons to retain this review:

Openness: Some participants may be reluctant to commit themselves to statements of issues and values until they "see where all this is going." In this case, it is advantageous to ask them to go along with the process on a trial basis, and let them know that when it concludes, they will have an opportunity to decide whether the outcome is satisfactory for them, whether they want to modify earlier statements, and so on. Often, when reluctant participants are assured that they are not making a premature commitment or being lured onto a one-way street, they become open to being won over by the positive dynamics of the process.

Reassurance: For a decision to be ethically justified, it must be the outcome of an inclusive, transparent, and rigorous process. One aspect of a good outcome is that every participant feels they have been part of a decision that is the best possible one. What the review does, by going over the logic and the steps, is show everybody that they have done a good job.

Efficiency: Despite the importance of the decision, for the patient's quality of life and for significant resource allocations, there is a natural tendency during the process steps to want to hasten things along by jumping to a solution. People tend to want to move into operational details right from the beginning. And yet it is crucial that we take all these steps in order to achieve an ethically justified decision. The consult leader often needs to put the brakes on and keep attention focused on the step at hand. The fact that there is a thorough review at the end, when all the details are checked against earlier decisions, allows those decisions to be properly made.

Improvement: Sometimes modifications and corrections make a good solution better. When moving from general considerations to practical details, it is quite true that one does not really "know where all this is going" until you get there. Especially when participants have been hurried, or are particularly enthusiastic, they may have moved off the track a bit along the way. It always helps to place the proposed solution against the lists of issues, values, and options to make sure that it really does fulfill what, at an earlier stage, people thought was important. So the review is intended not only to reassure but also to "make sure," and to make changes that seem appropriate.

For many reasons, therefore, an ethically justified decision usually depends on a final, meaningful confirmation step. This step concludes the consult meeting and generates the essential outcome of the process. The involvement of participants in decision-making is now completed. However, for the consultant or consult team, there remain some important steps in the Post-consult stage, to which we now turn.

Resources in the Toolkit
The *Toolkit* provides specific directions and suggestions for preparing and conducting the group consult meeting, as well as a four-page *Consult Meeting Data Capture Form* which can be found on pages 26–29.

2.5 Consult Stage 5: Post-consult

There are four steps in this final stage: documentation, follow-up support plan, evaluation, and identification of systemic issues.

2.5.1 Documentation

Although documentation is mentioned in the post-consult stage, it should be happening throughout the consult process. There are two types of documentation that the consult team should be doing. The first concerns the articulation of key process steps and content areas to be shared with those involved in the situation. The second is a summary of the context and analysis internal to the consult team.

An important purpose of documentation is to assist those involved in the situation to understand the ethics consultation process broadly, as well as where the process is at, at a given time. Indicating in the patient's chart when a consult has been initiated, and what those involved can expect, can help demystify the consult process and prepare those involved to participate effectively. Depending on the duration of the consult, further notes explaining the work done to date and next steps can help those involved feel more confident about the consult process and, again, be prepared for necessary participation. At the end of the consult, it is similarly helpful to send a note indicating the consult has been closed.

The documentation helps to share relevant information with those involved in the situation who are not part of the ethics consult. Documentation will communicate relevant key facts and values, analysis, important decisions reached through the consult process, recommendations being made by the ethics consult service and the rationale for these.

An interesting question about documentation is how detailed it should be. Some argue for brief documentation because, in the busy world of health care, there is little time to actually read long reports. In my experience, well written notes that reach three to five single-spaced typewritten pages will be read by team members and can serve as effective tools for education. For those rare cases of very high conflict that have the potential to be exposed more broadly through media or legal action, carefully written consult notes can be very important for describing the process of ethics analysis in general and explaining the specific analysis of the issues in question to a much broader audience.

> **Resources in the Toolkit**
> The *Toolkit* provides a *Consult Documentation Form* on page 14 to guide the capturing and sharing of information.

The emerging story should also be documented throughout the consult process (the *Toolkit* provides an *Emerging Story Form* on page 34 to help support this practice). The purpose of this is for the consult team to consolidate the story as informed by multiple sources to assist with their thinking and planning. This becomes even more important if there are multiple members of the consult team. The emerging story should include:

- Relevant demographic details of the patient and family
- Relevant details of care providers and teams involved
- Sources consulted
- Interviews and meetings planned and held
- Perspectives of participants
- Elements of the decision process, from key questions, facts and values to analysis, decisions and rationale

2.5.2 Follow-up Support Plan

The second step in the post-consult stage is the follow-up support plan. In my view, it's important to recognize that those teams requesting an ethics consult may actually be ripe for ongoing support from an ethics service. So the goal of an ethics consultation service is not simply to jump in and put out fires. Rather, we want to be building broader capacity for dealing with ethically challenging situations throughout the organization. Therefore, a situation where the need for an ethics consult arises may indicate that the team is actually open to having a broader conversation about their values and beliefs, about tensions among team members, and about how the team might deal with ethical issues they regularly face. This follow-up step is aimed at providing ongoing support around ethics in care.

> **Resources in the Toolkit**
> The *Toolkit* provides a *Team Support Follow-up Form* on page 32 to support this step.

2.5.3 Evaluating the Ethics Consult

The third step in the post-consult is evaluation. Looking back over the process, we want to assess how the decision was understood by all involved, whether all concerns were made explicit in the discussion, whether the decision still seems the best option, whether the participants felt respected and assisted by the process, and whether the participants felt the process had a successful result and a well-made decision.

If not, what was the root of the dissatisfaction? Were the conclusions that were drawn temporary, tentative, or definite? If the decision was achieved by consensus, did anyone seem uncomfortable with the decision, or, if consensus was not achieved, was anyone unable to live with a majority opinion? And how much has this process furthered the education of the health care providers so that they are better enabled to manage future issues?

This evaluation may be done in various ways. The ethics team could review the consult with a decision-making framework at an ethics committee meeting following the consultation. They could create a review panel consisting of members from the ethics resource team who deal exclusively with evaluating consults. Further, they could set up policy that requires evaluations to occur within a time limit upon the completion of the consultation.

All of this suggests that there are some very important steps for committees to take. They need to focus on skill development for committee members. They need to develop and determine their model and philosophy of consultation. They need to be very clear about the ethics consultation service, especially with respect to three stages: pre-consult, consult meeting, and post-consult. Finally, it's very important that the consult service evaluate its practices on an ongoing basis.

Resources in the Toolkit
The *Toolkit* provides an *Evaluation Form* on page 31 to support this step.

2.5.4 Identifying Systemic Issues

The final step in the post-consult stage is the identification of systemic issues. The ethics consult team should reflect on:

- What gave rise upstream to the request for the consult?
- What could have been done upstream to prevent the issue from arising in the first place?
- Are there recommendations that we can make for changes upstream that may prevent the demand for these kinds of consults downstream over time?

Resources in the Toolkit
For the Post-consult stage, the *Toolkit* provides a number of useful forms for consultants, including:

– **Consult Documentation Form**
– **Evaluation Form**
– **Team Support Follow-up Form**
– **Systemic Analysis Form.**

2.6 Conclusion

Ethics consultation services are sophisticated support structures that have become an important part of the modern health system.

The process for clinical ethics consultation described in this guidebook can allow ethics consultants to support decision-making that is inclusive, respectful and fully deliberative in contexts of diversity and power inequity. The process is meant to be sensitive and hospitable to ambiguity, interpretation and context. Most importantly, the process seeks to make room for conversations about the good life and what it means to be a good human being as those involved arrive at what is the right thing to do.

Balancing considered judgments about what is important in actual situations with deeper commitments, the inductive process calls for allowing those involved to express what is at stake for them. This eventually leads to the articulation of value themes that capture what should matter according to all those involved in the situation.

It is my sincere hope that the set of conceptual and practical tools in this guidebook, along with the accompanying toolkit, will help will assist those participating in the practice of ethics consultation to critically reflect on and improve their practice. The real end of this work, of course, is to improve the quality of ethical justification for responses to challenging issues such that those affected can live with greater integrity.